THE HOUSEBUILDING EXPERIENCE

THE HOUSE-BUILDING EXPERIENCE

JACK McLAUGHLIN

VNR **VAN NOSTRAND REINHOLD COMPANY**
NEW YORK   CINCINNATI   TORONTO   LONDON   MELBOURNE

Portions of this book first appeared, in somewhat different form, in "Populist Housing in America: Terpitecture," by Jack McLaughlin, published in *The Humanist* May/June 1978. Reprinted by permission.

All illustrations are by the author unless otherwise noted.

Printed in the United States of America
Designed by Karin Batten

Published by Van Nostrand Reinhold Company
135 West 50th Street
New York, NY 10020

Van Nostrand Reinhold Limited
1410 Birchmount Road
Scarborough, Ontario MlP 2E7, Canada

Van Nostrand Reinhold Australia Pty. Ltd.
17 Queen Street
Mitcham, Victoria 3132, Australia

Van Nostrand Reinhold Company Limited
Molly Millars Lane
Wokingham, Berkshire, England

16 15 14 13 12 11 10 9 8 7 6 5 4 3 2 1

Library of Congress Cataloging in Publication Data
McLaughlin, Jack, date
   The housebuilding experience.

   Bibliography: p.
   Includes index.
   1. House construction—Amateurs' manuals.
   2. Self-help housing.   I.   Title.
TH4815.M28   690'.837   80-24670
ISBN 0-442-25398-2

CONTENTS

v

ACKNOWLEDGMENTS

This book would not have been possible without the numerous owner-builders throughout the country who shared their construction experiences with me. To these cooperative and generous men and women, and to the USDA County Extension Agents and building materials suppliers who led me to many of them, I acknowledge a debt of gratitude.

In particular, I want to thank Mark Steadman, who started me off on my own house-building journey by demonstrating that it could be done with grace and style. I also thank Norman and Lila Book; Dick and Lucy Manson; John and Debbie Johnson; Don Miller; Mimi and Jere Hodgin; Robert Marshall, Director of Self-Help Enterprises; Robert Roskind of The Owner Builder Center; Jerry Newman and Luther Godbey of the USDA Rural Housing Research Unit; and David A. Murdock, of the Department of Housing and Urban Development, all of whom were helpful in various ways. Norman Holland, of the State University of New York at Buffalo, supplied some of the theoretical underpinnings of this book through his work on subjectivity in literature and the arts. Barbara Ravage guided this book through its editorial labyrinths with intelligence and sensitivity; I am grateful to her for her generous help and encouragement.

Acknowledgment is made to *The Humanist* for permission to reprint copyrighted material from "Populist Housing in America: Terpitecture," by Jack McLaughlin, in the June, 1978, issue.

My deepest gratitude goes to my family: my stepdaughter Suzi, who gladly donned hard hat and carpenter's apron; my son Kevin, who learned the art of housebuilding with me; and my wife Joan—planner, helper, editor, proofreader. To her, this book is lovingly dedicated.

INTRODUCTION

terp·i·tec·ture (tĕr′pə tek′chər), *n*. The art
of building for pleasure, particularly one's
own dwelling. [Indo-Euro. *terp* to please
oneself + Gr. *tektonikos* good builder]

Each year in America approximately one-quarter of a million
men, women, and their families take upon themselves the responsibility for constructing their own shelter. This represents
roughly one of every six single-family houses built annually in
this country—a remarkable figure that has remained stable for
at least the past decade. Not all of these amateur builders
complete the entire house themselves, but all of them assume
the planning, management, and control of the construction
process. This autonomy defines the owner as the builder.

Since each of these owner-builders acts independently, their
endeavors largely elude the statistical nets that capture most
of the events of our lives. The Census Bureau keeps a record
of the raw demographics—the number of owner-builders, the
size of their houses, the number of bedrooms and baths, the
kind of heating they install, and the method used to finance
their houses. But these tell nothing about the crucial facts that
differentiate a self-built house from a purchased house. How
well was it constructed? Was it difficult to build? How long did
it take? Who helped build it? How much money was saved?
How hazardous is building a house—can you get hurt? How
much skill does it take? Could I do it, and if I did, would I
be able to finish it? Would I enjoy doing it? The answers to
these kinds of personal, subjective questions are not to be found
in either the statistics available or in any of the numerous
housebuilding manuals. Yet they are the truly important questions facing those interested in the owner-builder movement

An owner-built house from another era, showing the builder's skills—handsomely chinked log walls, an arched stone carport, and sheet-metal roofing. Homes like this represent a populist craft movement that has been a part of American culture for generations.

as a sociological phenomenon as well as those who are considering building a house themselves.

In an effort to get some answers to these motivational questions I queried hundreds of owner-builders from every section of the nation by questionnaire and interview. To the collected wisdom of these owner-builders I've added my own experience in building my own house from foundation to rooftop. Together this represents, I believe, some of the most accurate, comprehensive, and authoritative information available on how Americans build their own houses and make the experience a deeply felt personal accomplishment.

The reader will notice that this book has few illustrations of house foundations, framing, masonry, carpentry; or diagrams of plumbing or electrical work—the things one finds in how-to-build-a-house books. There are many excellent housebuilding manuals which show in detail how a house is put together, and

I have no intention of adding to them. (There is, however, an annotated bibliography of some of these works at the end of the text. Information on the availability of books and publications cited in the text can also be found in the Bibliography.) This book can be read as a companion to these manuals, for it deals with housebuilding as a human, creative act, and illustrates how designing and constructing one's own shelter can be one of the most rewarding, pleasureful, self-actualizing events of a lifetime.

I've coined a term to describe this behavior: terpitecture. The word is derived from the Indo-European root *terp*, meaning "to please oneself" (the muse Terpsichore danced for sheer delight) and *tektonikos*, Greek for "a good builder." The terpitect builds his or her own shelter for pleasure. Building one's own house requires a number of specific skills and various kinds of technical knowledge, but mastering these does not guarantee that housebuilding will be a gratifying experience. Constructing one's own shelter is a populist art form, not a science. No two owner-built houses are ever exactly the same. The plan selected, the materials used, the work methods employed, how much of the construction is done oneself, how much time it

A contemporary owner-built house that is indistinguishable from purchased homes in the same community. Mark Steadman, the novelist, built this commodious, 5,700-square-foot house for his large family.

takes—all are unique to each builder, and all are the product of literally thousands of private choices. Building a house is really a phenomenology of choice, and how owner-builders make these choices will dictate whether they are engaged in construction or terpitecture.

One thing that I have learned from my research on owner-built houses is that any dogmatic statement about how owner-builders act or should act is almost certainly going to be wrong. The only statement one can make with certainty is that terpitects *do* manage to get their houses completed, and they find the experience richly satisfying. The ways they conduct the building process, the materials they use, the labor resources they assemble—all of these are as idiosyncratic as the personalities of the individuals involved. One can only marvel at the plurality of owner-building and admire the richness and ingenuity of the terpitect's creative resources.

There is a temptation by those who write about the owner-builder movement to impose their own standards, aesthetic and functional, on the would-be owner-builder. Although I have been aware of this danger, I have no doubt been guilty of the same kind of subjectivity; I have assumed that what worked for me is best, that my solutions are the Platonic forms of self-help building. There is no archetypical terpitect, however; I describe in Chapter 1, for example, how owner-builders re-create their own identities in their houses, and how each building enterprise is as unique as its creator. Subsequent chapters explore the economics of self-help houses—how they are financed and how they often get built without long-term debt. There are sections on skills—how easy or difficult it is for amateurs to construct various building systems, and how they go about learning them. The choice of building materials, the use of tools, and the organization of work are discussed as practical considerations, but I have also written about less tangible, equally important matters: the creative conflicts and pleasures of designing and building one's own home; the psychological resources assembled for a successful building experience; optimal use of paid workers; the cost of free help; and family relations while the house is being built.

Not all owner-builders construct from the ground up; in the

inner cities tens of thousands of individuals and families are homesteading, rehabilitating, and restoring run-down and abandoned apartments and houses. I have investigated the economics, sociology, and psychology of homesteading and rehabbing, and the role government plays and ought to play in these projects. Finally, I offer a prospectus for a national association of owner-builders, an organization that could provide terpitects with the social and political power they now lack.

In describing the multifarious ways Americans create their own shelter, I have tried to offer some hope for the hundreds of thousands of men and women who, because of ever-increasing inflation, see the possibility of owning a home slipping helplessly beyond their means. Although I do not believe that saving money is the most satisfying motivation for building your own house, it enters prominently into every owner-builder's personal equation. The "sweat equity" of a person's own labor helps many owner-builders construct a house that otherwise they could not afford. This book will demonstrate that building a house is not a task limited to a privileged caste of talented craftspeople. Those who build their own houses are for the most part quite ordinary citizens from all occupations, from all economic classes, with the same diversity of interests and backgrounds found in any cross-section of the population. What they share is the old-fashioned Emersonian virtue of self-reliance, the belief—sometimes the burning conviction—that you can, must, take into your own hands the task of providing shelter for yourself and your family. If this belief in personal autonomy is held strongly enough a way is found, obstacles fall away, skills are mastered, finances arranged, objections overcome. The vision of a dream house will become a reality and the dreamer metamorphosed into a terpitect.

THE HOUSEBUILDING EXPERIENCE

ONE

MY HOUSE MYSELF:
Planning and Designing the Owner-Built Dwelling

Let us begin with an axiom: in designing and building a house, owner-builders re-create themselves. Coming to terms with this simple idea will explain many of the puzzling decisions that are made during planning and construction and also will point out the origins of a good number of joys and conflicts. What it means is that the way owner-builders think about and, more importantly, feel about their building enterprise is a function of their unique identities. This fusion of object with maker is readily admitted in artists—writers, sculptors, musicians, painters—but it is not often recognized in a popular craft such as self-help building. This is possibly because it rarely exists in professional construction workers; they produce carpentry, masonry, electrical wiring, or plumbing for wages. Their aim is to produce a paycheck, not a creatively satisfying house system that mirrors its maker.

1

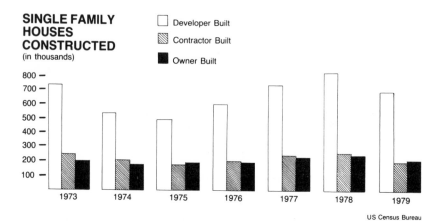

The number of owner-built houses has remained stable through the years in spite of the ups and downs of the new housing market. Housing recessions and booms cause wide variations in the number of houses built for sale, but owners who build their own homes, like those who have homes custom-built for them by contractors, are much less affected by current economic conditions than are those who purchase a home.

Nor does their work involve challenge and mastery; it is the skilled but routine response of the professional, repeated endlessly and mechanically.

Owner-builders are creative precisely because they are amateurs in the original meaning of the word: those who love what they are doing. Every decision made; every skill mastered; every piece of construction material measured, cut, or nailed; each error made; each success recorded—all are shaped by a unique pattern of desires, needs, and defenses. These habitual ways of coping with reality are reflected in every choice and are therefore an intimate part of the building venture. If one could read these decisions, one would find a blueprint of the builder. Terpitects create a house, and the house re-creates the terpitect. They are what they build.

This means that each decision owner-builders make is right for them, no matter how conventional or unconventional it may appear to an outside observer. This is, of course, an oversimplification because most owner-built houses are family projects with husband/wife/children decisions that involve numerous compromises. Yet as each house system is planned and built, one member of the owner-builder team usually stamps his or her personal style upon it. Many of the conflicts between couples who are builders occur because each is attempting to imprint upon the house his or her personality. One, for example, may feel the need for an open, expansive floor plan encouraging integrative family activities, while the other wants numerous closed spaces offering privacy and security. Realizing that these needs are not capricious or arbitrary, but are the very scaffolding of a person's identity, can smooth the way for successful compromises.

Nowhere is it more important to recognize that home and self are one than in the planning stages of the project, for it is here that decisions are cast in concrete and are not easily altered. During the actual construction of the house it is usually possible to feel out the family's real desires through trial and error. Changes may be awkward and expensive, but they are seldom impossible. But the planning decisions often scribe the ultimate shape of the house with such an indelible line that it cannot be erased. Builders must take particular care, then, that their plans are chosen with *their* goals in mind and are not forced upon them by either the will of outsiders—including myself—or by social conventions. As for the latter, terpitects must always remember that conventional wisdom—particularly from bankers and real estate agents—has it that new houses are purchased from an assembly line, like cars and tractors, and that only lunatics would contemplate actually building one themselves.

DESIGNING FOR IDENTITY

Every owner-builder begins with an idea, often quite vague, of the kind of house desired. The genesis of this idea is rarely known; if it were to be traced back through memory perhaps its origins could be located. Perhaps an "ideal" house is one like the house you grew up in or visited as a child. Maybe it was a friend's or relative's house that holds *les recherches du temps perdu.* It may be a house that was seen only at a glance, but which represented a desired style of life, upward mobility, a fashionable neighborhood, an escape from an oppressive past. Or it may be precisely like every other house in the neighborhood, and therefore represents a feeling of social cohesion and belonging. The current manifestation of this ideal house may be a photograph in a home magazine or a house driven past quite by accident, but if you ask what it is that this house is a reminder of, it may be that the answer will touch upon the very quick of your personality.

Realizing that a house is a kind of *Doppelgänger*—identity translated into shaped space—leads to the kinds of questions that will produce a satisfying design. Such an analysis begins by asking about interpersonal relationships with family, friends, and acquaintances, then moves on to individual preferences:

Do I (or we) want open living areas where people can interact freely, where play, conversation, and eating are family rituals, or do I want controlled zones where I can place limits on social interaction as my mood dictates?

Do I want a house with formal living spaces tastefully decorated and fastidiously ordered, or do I enjoy living in loosely controlled chaos, in a carelessly arranged house that looks "lived in"?

How important is food? Is the kitchen merely a machine for preparing meals, or is it a social space, a hobby space, a creative space?

How important is bathing? Do I take quick showers or long, slow soaks? Do I read in the bathroom? Is the bath a sexual place? a greenhouse?

Do I conceive of the master bedroom as a special place, spacious, luxurious, playful, erotic—or merely a sleeping closet? How far away should it be from the other bedrooms?

Do I prefer to have the outdoors stay outside, or would I like to keep the dividers between inside and outside flexible and ambiguous? How do I feel about glass?

Do I like parties to be big bashes with guests circulating through the house like gypsy hordes, or do I prefer intimate, candlelit dinner parties with a few friends? or both? How do I feel about formal dining rooms? wasted space? absolutely essential?

Is a nursery in the foreseeable future? Where do babies belong in a house? young children?

Is television worth a special room, or is it a nuisance to be hidden from view?

How important are hobbies in my life: books, sewing, shop work, cards or games, cars, boats, sports? Is the hobby or its equipment worth a special space?

Is a garage a place for cars or a large storage area? Would a carport do?

These may seem to be the kinds of questions that only an architect would ask a client; but owner-builders are their own

The same basic floor plan can be adapted to different personalities. The plan on the left reflects a desire for closed, well-defined spaces. Rooms can be closed for privacy and solitude. The plan on the right is much more open, with loosely defined spaces, allowing for much more interpersonal contact.

architects, and they have the freedom to design any ordering of space they desire. Floor plan ideas usually come from home magazines, house plan magazines, or friends' homes. In designing a floor plan, the owner-builder is free to borrow from any source, to adapt any plan to his or her own needs. Families who are building their own homes often have members make a list of the kinds of spaces each would like so compromises can be negotiated.

Owner-builders who are constructing a house for the first time may feel that building a house, when they know nothing about how to do it, is difficult enough without designing it also. My survey of owner-builders, however, demonstrates that designing the house is the one act that most self-help builders share. Of my sample, 60 percent designed their houses themselves, and 20 percent bought blueprints from a plan service. But most of those who purchased plans, including myself, altered the plans considerably to satisfy their own needs. The 5 percent who used an architect or engineer generally designed their own houses and used the professional merely to draw up sets of blueprints or engineering specifications. An additional 5 percent copied another house, but this too incorporated numerous alterations. Even the 10 percent who built precut or package homes altered the package plan to some degree and finished the interior of the house according to their own designs.

The original design of the house does not tell the full story of how owner-builders re-create themselves in their homes. An owner-built house is an ongoing process rather than a finished artifact. During the construction process almost every owner-builder I know made some changes in the original concept of the house—some major, others refinements of earlier designs. No matter how well a person can conceptualize three-dimensional space from a two-dimensional blueprint, no one really knows what a space *feels* like until it is constructed and inhabited. After the house is framed and roofed, and the owner-builder is able to stroll through the floor plan for the first time, there are often second thoughts about how spaces are arranged, of how wall partitions feel, of the size of windows, the placement of kitchen and bath cabinets. It is rarely that a change of mind cannot be incorporated into the house during construction. In my own case, once the frame of the house was up, I extem-

porized the interior spaces of the house. Post-and-beam construction, which requires no load-carrying walls, allowed me
complete freedom to design as I built. The house I have now
evolved over the three-year period it took to complete it, and
I believe that that evolution helped the house reflect my family's
personality more than if its design had remained inflexible from
the start of the construction project.

The kinds of owner-built houses described in magazine and
newspaper articles give a false impression of the design styles
most self-help builders choose. These well-publicized houses
are likely to be offbeat or innovative: the young architect's
owner-built dream house, the "alternate life style" A-frame in
the woods, farmhouses built for next to nothing, underground
homes, domes, yurts, log cabins. Such dwellings give an idea
of the range, imagination, and diversity of terpitecture, as well
as the creativity and individuality that define it, but they are
the interesting exceptions rather than the rule. The spine of
the owner-builder movement is the three-bedroom, two-bath
ranch or colonial house that is indistinguishable on the outside
from thousands of houses very much like it.

Most owner-built homes are indistinguishable from other houses in their
neighborhoods. The one-story ranch house, like this one, is one of the
most popular, easily constructed designs.

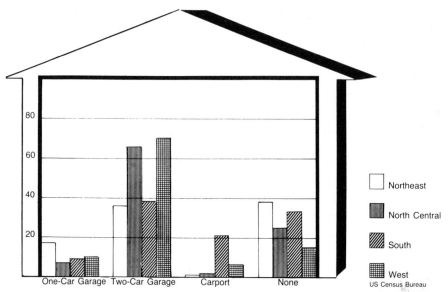

PERCENTAGE OF OWNER-BUILT HOMES WITH CARPORT OR GARAGE

Northeast

North Central

South

West

US Census Bureau

Geographical differences dictate a number of building practices. A two-car garage is the standard for most owner-built homes as well as for developer- and contractor-constructed houses. The South, however, is the only section of the country where carports are constructed in significant numbers. In the West all but 15 percent of owner-built homes have some kind of parking shelter.

My sample of owner-builders shows that roughly 70 percent construct traditional houses of one kind or another—ranch, split level, colonial—and about 30 percent build contemporary houses. What this means is that owner-builders are for the most part middle-class Americans who share middle-class American values. The style of their houses, the face shown to the outside world, is not very different from that of their house-buying neighbors. The differences are in the adaptation of that style to the builder's individuality and the way it is built.

DESIGNING FOR EASE OF BUILDING

There are design decisions that are unique to the owner-builder. These have to do with the construction process, not with the end product. Those who purchase homes from contractors or developers sometimes design the floor plan, but the problems of how the plan is translated into a finished house is the concern of the professionals who build it. This is not the case with owner-builders; each of their design decisions are inextricably bound up with construction decisions. They must not only design, but execute, which means that they can design only what they feel they are able to build or supervise others to build. If they plan to subcontract all of the building systems, this is less of a problem (but see Chapter 8 on difficulties with subcontractors); however, most builders do some of the work themselves.

One of the first decisions owner-builders must make is whether they will construct a one-story or two-story house. A two-story house has much to recommend it; the second story provides extremely inexpensive space because one roof covers two floors. It uses less material per square foot of floor space than a one-story house, and is also more economically heated. Plumbing can be minimal if baths are stacked on top of each other, and the second story can be roughed in and remain unfinished until more time or money is available. But it is not as easy to build as a one-story house, if only because much of the work is performed twenty or thirty feet from the ground. This means that scaffolding and ladders are necessary for many jobs, and materials for the second floor must be lifted or hauled to considerable heights. Two-story designs, such as the Cape Cod house, also have steep, treacherous roofs; and masonry work on chimneys—one of the more difficult, hazardous, and expensive house systems—is doubled with a two-story house.

A one-story house, on the other hand, keeps the builder much more comfortably close to the earth; there is less need for scaffolding and for working on ladders, and there is much less risk of injury. This may not seem too crucial, and to the professional who has the necessary scaffolding and labor, it is not. But for the owner-builder working alone or with insufficient help, heights present a very real problem—in getting materials up there and on working precariously and often dangerously. The one-story house, however, is not as economical in its use of materials, and its increased wall and roof areas account for greater heat losses. Broken roof lines can add additional expense and difficulty during construction, and separated kitchen, laundry room, and baths require more complex plumbing and wiring systems.

Census Bureau figures show that over the years, owner-builders were constructing three one-story houses for each two-story house; but in recent years, possibly because of the increased costs of energy, more two-story houses are being designed and built. There are sharp geographical preferences in this: in the Northeast, 60 percent of the owner-built houses are two-story (this is no doubt due to the colonial and Cape Cod tradition, as well as climate) whereas in the South, 80 percent are one-story. These statistics demonstrate the role cultural conformity plays in decisions about shelter. Owner-builders are no different from their home-purchasing neighbors in this respect: they tend to build the types of houses that already exist in their community.

ECONOMY AND SIMPLICITY OF DESIGN

Rarely do political events influence the design of individual shelter; changes evolve from shifts in cultural values, and usually occur slowly. A notable exception was the Arab oil boycott of 1973. Before the meteoric rise in the costs of energy, insulation was of minor importance in house design; heat loss through large expanses of glass was inconsequential; the size and number of rooms had little to do with heating or cooling costs. All that, of course, has changed. Energy costs are now a major consideration in house style and construction design. In many areas of the country the costs of heating and cooling a house are higher than mortgage payments.

Most of the owner-builders I queried started their homes before the energy crunch, but all admitted that they would have designed for more energy efficiency if they were building today. For example, post-and-beam construction has been, and continues to be, one of the most popular architectural styles in the nation—not only in contemporary houses, but also where it is applicable in more traditional styles. Many of the architect-designed houses featured in the home magazines are built this way, with large expanses of glass, and towering cathedral ceilings. My own house is this style, but if I were designing it today I doubt that I would choose to build it this way. The south side of the house is virtually a glass wall, and the plank-and-beam ceilings in all of the rooms simply did not allow the amount of rigid insulation that I would now like to have. Cathedral ceilings are marvelous in the summer when hot air rises and living areas are cooler, but if you have lofts or bedrooms opening onto the upper areas they can be impossible to cool on hot days. In the winter, the upper level is the warmest place in the house, but the living areas are invariably cold and drafty at night. (However, in passive solar design, discussed in a later section, these open styles can be a distinct advantage.) In the sunbelt area of the country where I live, these problems are less acute, but in the North and Midwest they are major considerations.

For the owner-builder, house size has always been critical. Over the years, self-built houses have been smaller on the average than purchased houses for several good reasons. Economy dictates a smaller house for many home builders; they build what they can afford. This is particularly true of rural housing, and often true of young builders who are constructing their first house. Time and personal resources are also important considerations. A builder designing his or her home must always take into account what can be accomplished in a reasonable amount of time. Few builders can enjoy the luxury of an extended period of time off from regular work to construct a house; it must be completed in spare hours. A large house, and the increased amount of time it takes to finish it, often seems too much of a project to tackle. When faced with the realities of building, owner-builders frequently scale down the size of the house and settle for an open-ended design that can be expanded in the future as a larger family or a different life style dictates.

With the new energy picture, the size of a house is also closely related to maintenance expenses: the larger the house, the greater the cost to heat and cool it. Most Americans have grown up accepting the axiom that bigger is better, but the accelerating costs of energy have forced most of us to question that belief (the automotive industry has been coerced by the government into abandoning it). With thoughtful design it is possible to give the illusion of size while controlling the actual square footage that must be heated and cooled. For example, one of the chief benefits of the trussed gable roof is that, aside from the mechanical kitchen-bath cores, there are no necessary, fixed interior walls. Owner-builders have complete freedom to control the interior spaces as they like; partial walls can be placed where they will open up space to create the illusion of size, and closets and utility rooms can be partitioned off where they are needed.

Not to be overlooked in any discussion of economy of maintenance is the fact that designing for such economies is itself a source of gratification. As I discuss in Chapter 2, getting a bargain is one of life's universal pleasures. Builders who design energy-efficient homes produce an infinite source of delight: every month they receive their bill from the electric company, oil distributor, or gas company they will feel the rush of pride that comes from a bargain well struck. Terpitects design economically and in doing so assure a continuing bounty of self-satisfaction.

Simplicity of design is particularly valuable for the novice owner-builder. Ken Kern, one of the most experienced and wisest commentators on the owner-builder movement (see his *The Owner-Built Home*) made the comment that one of the main reasons some builders fail to achieve either personal or aesthetic satisfaction from their projects is that they choose to build houses that are too complex. "I have always been at a loss," he remarked, "to determine why it is that unskilled owner-builders invariably attempt to build complicated structures. The less ability they have to draw upon, the more structurally involved the project becomes." An owner-builder's design objective might well be to weigh the value of increased space and complexity of building systems against what can realistically be completed in a reasonable amount of time and at a comfortable cost. But even after writing this, I must hastily add that complexity may be a way of life for some owner-builders—and in this case, a complex house, with all the difficulties it entails, may be absolutely right.

THE OPTIMUM VALUE ENGINEERING SYSTEM

Before any owner-builder makes a final decision on house design—one-story versus two-story, kind of foundation and roof desired, dimensions of the house—he or she might first examine the Optimum Value Engineering (OVE) System. This is a series of building techniques developed by the National Association of Home Builders' Research Foundation in cooperation with the U.S. Department of Housing and Urban Development, and it is designed to enable builders to construct single-family houses with maximum cost efficiency. Although the OVE system was created for commercial home builders, it is ideally suited to the owner-builder who is interested in designing for ease of construction and economy.

The OVE manual, *Reducing Home Building Costs With Optimum Value Engineered Design and Construction*, should be one of the first books that any owner-builder purchases. It is literally possible to design a house from foundation to interior trim by following the construction details outlined in this manual. The main value of the OVE system for owner-builders, however, is that it describes a set of planning, engineering, and building practices that will take much of the mystery out of house framing, reduce materials to the absolute minimum consistent with safety, and give builders the maximum amount of flexibility in shaping house to personality.

These are some of the highlights of the OVE system:

Design a house based on a two-foot module, or even better, a four-foot module. Since all building materials are cut in some multiples of two, and plywood in multiples of four, a modular house eliminates waste. Framing members in the floor, wall, and roof are also coordinated to a two-foot modular dimension.

If at all possible, design a rectangular house. This shape provides for the most economical use of floor and wall areas, limits the plan to four straight walls and simple roof structures, and still allows design flexibility. An optimal size for a house of 1,232 square feet, for example, would be twenty-eight by forty-four, a multiple of four feet in both directions. A garage or carport could be added for an L-shape without altering the rectangular plan of the house.

Design kitchen, baths, laundry, and heating/air-conditioning systems around a central core. This is the most efficient way of handling these mechanical systems and reduces material costs for plumbing and duct work.

Plan house framing twenty-four inches on center instead of the traditional sixteen inches. Most codes now permit spacing studs, joists and roof members at this distance; the savings in material is considerable. This spacing of studs also allows for windows that fit between the studs, eliminating headers and jack studs.

Use lightweight roof trusses. They can either be purchased from a truss fabricator, designed to exact specifications, or built on a jig laid out on the foundation floor. They allow builders to cover the house quickly so that it is out of the weather and dry. Because trusses are supported only by the outside walls they rest on, they give the owner-builder complete design flexibility; all interior walls in a truss-roof house are non-load-bear-

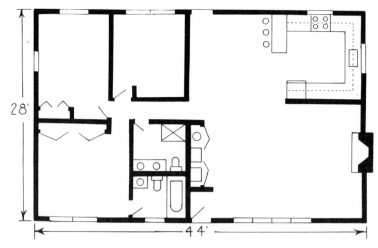

A modular floor plan such as this one, designed in multiples of four feet, produces maximum efficiency in the use of materials and in ease of building. A simple rectangular house is one of the easiest to build.

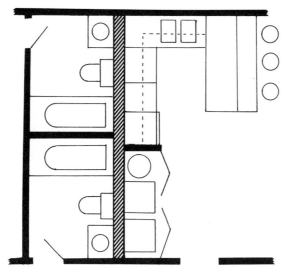

A central utility core places all plumbing and heating/air-conditioning in this floor plan in a single foot-wide wall. Plumbing runs and heating/AC ducts are minimized for cost efficiency and ease of building.

ing and can be placed anywhere the builder needs them. The OVE manual gives enough specifications for self-help builders to design and construct their own trusses.

Of the possible roof designs, a ⁴/12 gable roof (four inches of drop for every horizontal twelve inches) is the simplest to build and the easiest to work on and maintain. Water does not easily blow under the shingles, snow loads are usually not a problem except in heavy snow areas, and shingles or shakes can be applied quickly. Steep roofs, particularly on two-story houses, are extremely hazardous to work on. Flat and shed roofs are simple to build but are often costly to maintain because of the possibility of leaks. Also, finishing a built-up flat roof is usually a job for a professional crew; owner-builders can apply asphalt shingles to gable roofs themselves.

Do not overengineer a house. The OVE manual describes dozens of places where traditional materials can be eliminated. These suggestions are all based on engineering tests of reduced systems. Such new techniques as glue-nailing produce a stronger house with fewer materials.

Prefabricated roof trusses are a boon for the owner-builder. They make it possible to erect a finished roof quickly and to span a house width without support. Here, a team of Self-Help Enterprise builders lifts trusses into position. (Photo: George Ballis)

This last piece of advice, building with less, is a controversial idea. Many owner-builders feel that one reason so much commercially-built housing is shoddy is that it is thrown together with no concern for quality materials. Skimping on materials in one's own house seems to be doing the very thing that the owner-builder is attempting to avoid. I have mixed thoughts on this. My own house is unconscionably overbuilt; I estimate that I probably used at least 10 percent more building materials than the house required. This came about largely from not being sure of what I was doing, from not knowing for certain whether a system was structurally sound the way I was building it, or from simply not knowing the most efficient way to design a system. My house, therefore, cost me more than it should have—in time and money. On the other hand, it is very solidly built and I do not feel at all bad about that extra building material in it. But if I were building now, I would be more efficient. The house will go up much more quickly with fewer materials and simplified framing, and for an owner-builder, time is sometimes as valuable as building materials.

Even when owner-builders do not adopt the entire OVE system, they are certain to find that its recommendations are worth consideration. Those parts of the system that are incorporated into the design plan will more than repay the builder in efficiency and economy.

SOLAR ENERGY AND THE OWNER-BUILDER

Solar energy is on every home builder's mind these days, and the owner-builder is no exception. It is one of those ideas that seems so right—an abundant source of clean, free energy that is there for the taking. Everyone is aware of the single flaw in sunlight as an energy source: it is gathered with relative ease, but stored with great difficulty. In solar house design the collection-storage is handled by two basic methods: passive systems and active, or high-technology, systems.

The passive system is closely integrated into the design and construction of the house, and uses the sun to warm house materials directly without mechanical assistance. Basically, it exploits the "greenhouse effect" of glass to collect heat; and it uses massive thermal sinks, usually masonry but also rock and water, to store it. South-facing windows allow short-wave radiation to pass through the glass; these rays strike the interior surfaces of the house which, in turn, emit long-wave radiation that does not pass back through the glass. This produces a heat gain. Heavy masonry walls or floors have the capacity to store heat collected through glass for release at night or cloudy days. Passive systems operate entirely by the natural flow of heat by convection and radiation, although the air flow may occasionally be aided by small fans. One of the most common ways of storing heat in passive systems is the Trombe wall, named after one of its developers, French physicist Felix Trombe. This is a

THE SOLAR ATTIC HOUSE: A SOLAR SYSTEM FOR THE OWNER-BUILDER

The U.S. Department of Agriculture Rural Housing Institute has designed a solar house that is particularly well suited for owner-builder construction. The house collects heat in the attic and circulates it through a crushed rock storage unit in the crawl space. It is a simple, low-cost solar system that has been thoroughly tested and can be adapted to any ranch-style house. All it requires is three ducts running from the attic to the crawl space. Depending on the climate, it will supply from 60 to 80 percent of the house's heating needs.

Plans for the house, no. 7220, can be obtained free, or for a nominal charge, by writing to the Agricultural Engineering Extension Office of your state land grant university. A list of land grant universities can be found in the Bibliography.

THREE – BEDROOM, SOLAR – HEATED HOUSE   PLAN NO. **7220**

Living Area -- .1120 sq. ft.
Garage -- 336 sq. ft.
Patio -- 160 sq. ft.

The solar attic system is a flexible design suitable for a wide variety of floor plans. The solar attic collector itself is the essence of simplicity. Blueprints are available free or for a nominal fee through state land grant university agricultural engineering extension offices. (Exterior photo and plan: USDA Rural Housing Institute)

High energy costs have made glass walls such as these in the author's home virtually obsolete. Although the windows face south and collect ample quantities of winter sunshine, it is quickly lost through the clerestory glass at night. Heavy masonry walls and floors, combined with insulating shades and double glazing, would make a glass wall such as this much more energy efficient.

thick, interior masonry wall painted black on the southern, sunlit surface for maximum heat absorption and faced by double-glazed windows. Although passive solar systems can be highly sophisticated in their engineering, any newly constructed house that is oriented on an east-west axis with large, double-glazed windows facing south can utilize passive solar energy to some degree. A well-insulated house is absolutely essential, and the more heavy, interior masonry to absorb heat, the more effective the heat gain.

The active system collects heat in a series of flat plate collectors—metal and glass panels which trap heat and transfer it in either air or liquid form to a storage sink. It is then distributed throughout the house. This method often employs complex mechanical subsystems such as pumps, fans, sensors, thermostats, ducts, or plumbing; hence it is often referred to as a high-technology system. The collectors are usually mounted on a south-facing roof and the storage unit of rock or water is in the basement or crawl space. The distribution systems are widely varied, but they all include some kind of conventional back-up heater, because even highly efficient active solar systems rarely provide more than 50 to 70 percent of the heat required for an average home.

Since solar heating is relatively new, only a few of the owner-builders I surveyed had designed and constructed their own solar plants, although many of the experimental houses described in the vast literature on the subject are owner-built. I was fortunate enough, however, to be able to observe the construction of two solar houses built by a commercial solar construction firm for the USDA Rural Housing Research Unit. One was a one-story house with air-system roof collectors, gravel storage in the crawl space, and a solar greenhouse. The other was a one-story house, using a similar air system, but featuring earth berm construction—a mound of earth abuts three sides of the house.

As I watched these two houses go up I asked myself whether I, who had built my own house, could have constructed one of these and, if so, would I have wanted to. The answer to both questions was no. One of the problems with high-tech systems such as these is that at the present time owner-builders cannot buy the equipment off the shelf, as they can for a conventional

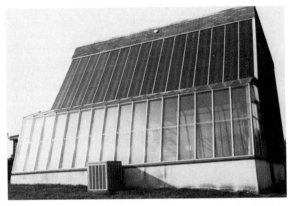

A solar greenhouse, such as this one attached to an active air-system solar house, provides growing space for plants and sunspace for people and is designed as a passive solar collector.

heating system. One can completely install an oil, gas, or electric forced-air system with hardware bought from local building supply stores. Solar equipment must be engineered and designed for each house, and the hardware is by no means easy to come by. Nor are the skills required simple ones. Even though I designed and installed the ductwork for the heating system in my own house, I was overwhelmed by the sheet metal work in these two houses, as well as by the complex series of dampers and servo-controls. I felt much more positively about a much simpler system—the solar attic house— also developed by the USDA Rural Housing Research Unit. I watched a successful version of this house constructed, with no particular difficulty, by an owner-builder. (See p. 18.)

My negative feelings about active air-system houses came from a conviction that novice owner-builders could very easily get in over their heads with the design, engineering, and construction of them. Building a conventional house is in itself a formidable challenge to the novice; to add to this the uncertainties and insecurities of dealing with experimental technology seems to me to be unwise. Furthermore, since so much depends on climate and construction efficiency, there are few guarantees that solar equipment will perform as expected. Building one's own house should be an act of creativity and joy; undertaking a house system with a high risk of frustration and anxiety seems to be an adventure for only the most daring and experienced builders. Of course, truly inventive terpitects will ignore everything I, or anyone else, says about active systems and build whatever inspires them anyway.

Subcontracting an active solar system is fraught with even more danger. It is currently outrageously expensive and there are not too many reliable and knowledgeable installers in the business. Anyone who contemplates hiring a contractor should consult Malcolm Wells and Irwin Spetgang's *How to Buy Solar Heating Without Getting Burnt*. In a survey of 100 owners of solar homes, the authors document the poor design, inferior workmanship, and high costs of many contractor-installed high-tech systems, as well as the unreliability of state-of-the-art hardware. Well-designed and reasonably-priced high-tech systems will no doubt appear on the market in the not-too-distant future, but in the meantime I would be inclined to hold off on such a system until I see the hardware, complete with instructions, in the latest Sears or Ward catalogs.

On the other hand, I do not think that anyone starting to design a self-built house should fail to consider solar energy. If active solar systems are currently questionable for the owner-builder, passive systems are worth serious consideration. The two crucial factors for any solar system—active or passive—are adequate sunlight and proper insulation. By orienting a house so that major glazed areas face south; by limiting windows on north, east, and west walls; by constructing overhangs to allow for full sunlight through double-glazed areas in winter and complete shade in summer; by providing for the free flow of air in interior spaces; by including insulated masonry walls or floors in south-facing rooms; by building with six-inch studs to allow for six inches of wall insulation—any or all of these building principles will lead to solar efficiency in an owner-built house. And the house need not have a single piece of "solar equipment."

It is also possible to design a roof that can be retrofitted with solar panels in the future, when more efficient, dependable, and inexpensive active systems come on the market. The earth berm solar house built by the Research Unit used a trussed roof system that angled the collectors for maximum light capture. These trusses were nailed together on a jig that had been set up on the subfloor of the house—something any owner-builder could do. If I were planning a one-story, ranch-style house I would certainly consider designing into it a similar truss system with the best roof pitch for my part of the country. Then I would make an allowance for space to incorporate a rock or water storage system and a way to get the rocks or water containers into that space. Such a system could also accommodate a water-heating unit in the future. And I would certainly include a solar greenhouse to provide heat, food, and beauty.

Jerry Newman, Director of the USDA Rural Housing Institute, offered one piece of advice based on his experience with the two experimental solar houses the institute has built and monitored. "It's absolutely essential that you build a tight house," he said. "Small leaks in any solar system—poor insulation, uncaulked cracks, improperly installed ducts, poorly fitted dampers—these all rob the system of its efficiency and lead to disappointment." Owner-builders who are willing to take the time to build carefully and well should be able to construct a solar house equal to or superior to one built by professionals.

One important consideration in any decision on solar heating systems is the tax credit, allowed by the federal Energy Act, for the purchase of solar equipment. This credit can be an incentive to owner-builders contemplating solar heating, particularly if they plan to do the installation themselves. The tax credit amounts to an appreciable discount on the purchase of solar hardware. Most states have also legislated financial incentives, usually in the form of property tax exemptions or reductions for solar heating/cooling. Every owner-builder should review all federal and state solar legislation before making any decisions on solar systems. Several states now require allowances be made for future solar installation. (*The Solar Energy Handbook*, published by *Popular Science* magazine, is an inexpensive source for much of this information.)

As in every other house system, solar energy must take its place among the variables which compete for the owner-builder's reserves of time, money, skills, and human resources.

Terpitects will invariably design into their houses as much solar heating as they can fit into their unique construction equations. Active solar heating is presently relatively expensive to construct (but promises future savings in heating costs); it places heavy demands on time because it is not standardized; and its payoffs in pleasures of accomplishment are not assured. But it is new, challenging, exciting—just the kind of enterprise that interests those who would build their own houses in the first place. Potential owner-builders are advised to dip into the literature on solar heating (see the Bibliography), talk to anyone they can find who is doing it or who has done it, and design as much of it into their homes as they feel comfortable with.

PRECUT AND PACKAGE HOUSES

An increasing number of owner-builders are constructing precut, or package houses, a building system that offers several advantages for the first-time builder. If the number of building supply chains that have entered this field is any indication, it is one of the fastest growing businesses in the housing industry. A precut house supplies owner-builders with the materials needed to finish all or part of the house, cut to specified sizes, delivered to their building site—sometimes with a work crew to erect the basic house shell. In some cases, the building materials are not precut at a factory but are delivered in the required amounts at the time needed for each building system. The builder job-cuts them on the site. In both cases, the house package usually includes such factory-produced components as structural framing systems, roof trusses, door and window units, siding, bathroom fixtures, and kitchen cabinets. The purchasers of a precut house can usually choose how much of the house they want—from a simple shell with no plumbing, wiring, or interior finishing to a complete house, down to the trim, wallpaper, and carpeting.

All precut and package house manufacturers offer a line of models for the purchaser to choose from; but many of them specialize in only a single style: colonials, Swiss chalets, log houses, vacation homes, domes, or contemporaries. Most of them also allow the purchaser to alter floor plans, choose from a variety of exteriors—in short, to "customize" the basic package house. Package buyers may have a local contractor build all or part of the house, or they may purchase the package, act as their own contractor, and build or subcontract as they see fit. In this case, they are essentially no different from other owner-builders who design a house and purchase building materials themselves.

The advantages of the precut house for owner-builders are that it eliminates much of the worry, uncertainty, and decision making that accompanies an enterprise about which they know very little. They work from blueprints and plans that have been developed by professionals. They often have a set of building instructions to guide them past many of the errors into which most owner-builders bungle. Building costs are also clearly delineated. Builders usually pay for materials with cash or a construction loan before starting, and therefore have a good idea how much the finished house is going to cost. If materials are precut, the house should go up much faster, not only because of the labor saved in cutting and fitting, but also because the builder does not waste hours running incessantly from one building supplier to another.

The disadvantages of the precut house are closely connected to its advantages. The professional help that builders receive becomes rigidly final when the shipments of materials arrive; there can be no second thoughts, no reevaluations based on actually standing in a rough-framed room and seeing for the first time the bleuprint dimensions realized in personal space. There are no allowances for the evolution of interior space as the house takes shape—all possibilities open to owner-builders who design as they go. By eliminating the difficult choices and uncertainties, the package house also eliminates the pleasures that come from serendipity, from the happy accident and the creative solution.

Owner-builders who purchase a precut house are also locked into the kinds of materials supplied by the manufacturer. If they want better quality windows and doors or a particular brand of roofing material or a special touch of luxury in one of the rooms, they often do not have the option to get them.

They also may not be able to make the package house as energy-efficient as they would like because of structural designs over which they have no control. They may not always be able to anticipate these problems until the house begins to take shape.

The cost advantage of any owner-built house is for the most part in labor saved. When owner-builders purchase a precut house they are paying for what is largely unskilled labor used in cutting the building materials, labor they could easily do themselves. They are gaining, of course, ease of construction and time. But there are other costs passed on, those that any company incurs: administration, overhead, and profit. They are also compelled to purchase or finance the house in advance—they cannot build from out-of-pocket, or pay as they build. Package buyers may feel that the freedom from some of the anxieties of building and the professional help are well worth the costs; and they may very well be, since so much of the home-building enterprise is subjective. The precut system has produced many satisfied owner-builders, but it is not without its share of bothersome problems.

In my owner-builder survey I heard from a number of precut builders, talked to several of them about their experiences, and saw their finished homes. Most of them had some complaints about the service they received—delayed shipments of materials, poor workmanship by crews supplied by the packager, missing building supplies. Many of these difficulties are the same as those voiced by every owner-builder who has to contend with workers who have no direct concern with the building project. One builder had a truckload of building supplies dumped on his neighbor's lawn; another found that the "skilled" construction workers supplied by the packager were parolees from a nearby prison. One precut purchaser accounted his griefs with a sigh: "Our lumber was precut, but much of it wasn't accurate. None of the loads of materials—plumbing, roofing, etc.—was complete; something was always omitted: starter shingles, 15 feet of fascia, roof edge molding, porch posts, light fixtures. . . . They made good on everything, but getting replacements was a problem because of shipping delays. There was even a two-month delay in getting the plans drawn, corrected, and re-corrected." And this was a house purchased from one of the largest packagers in the nation. Another owner-builder reported that the amount of money he saved on his precut home was less than 10 percent of the cost of a contractor-built house, and he did not feel the experience was worth it.

In Chapter 8, I'll discuss in detail the problems owner-builders face when they must depend on others for their labor, but the kinds of complaints listed above are clearly the result of builders losing autonomy over their projects. The more control you have over every part of every system of the construction venture, the more success you are likely to have. There is no guarantee that a large, national organization will provide better service than a small, local one since most of them operate on a franchise basis. Perhaps the best advice to be given to a potential precut purchaser is: talk to a number of owner-builders—not contractors—who have erected one of the company's homes and judge for yourself how dependable the firm is. And keep your fingers crossed.

ACQUIRING LAND

During design and planning a house exists only on paper, but eventually it must be moved to *terra firma*. Owner-builders may already own the land and design the house to the particular contours of the site, or they may decide upon a house and then search for an appropriate piece of land to fit it. In either case, land and house should ideally meet, fall in love, and be wedded.

Land is not always purchased; sometimes it is acquired as a gift. The vanishing small American farm, for example, has supplied building sites for thousands of owner-built homes. Fathers, who once handed down a working farm to their sons, now supply raw land to build a house on. In my sample of owner-builders I found that about 10 percent were given the land for their houses, or the money to buy it, by their family. This figure might inspire would-be builders to make discreet inquiries if there is land, or generosity, in the family. It is being done, and it has tax advantages for the giver.

Most owner-builders in my survey, however, bought their land for cash; a lesser number took out a loan for it. In either case, buying land gives owner-builders the opportunity to choose where the home will rest. In selecting a building lot or acreage, the self-help builder shares with the home buyer certain considerations: the character of the neighborhood or community; property tax; zoning and building restrictions; availability of utilities and paved roads to the building site; closeness to schools, churches, markets, playgrounds, day nurseries, and cultural centers; distance from one's business or place of work; and availability of public transportation. But there are other considerations peculiar to the owner-builder. While it is desirable for anyone moving into a new community to have congenial, friendly neighbors, it is important for the owner-builder to have cooperative ones. It is disruptive for anyone to have a house built next door; construction is noisy and dirty, strangers come and go, often with little concern for the social amenities of the neighborhood; and a construction site is an attractive but dangerous place for children. It is bad enough for this to go on for several months, but when an owner-builder is constructing a house it can last for a year or two—maybe longer.

If neighbors are hostile to the owner-builder's extended project, they can be a constant source of anxiety and frustration and may turn an enjoyable venture sour. Of course, the builder can take the attitude that it's my land and I can do what I like on it. But this is not quite true, for he or she depends at very least upon the passive cooperation of neighbors—that they will not do anything to disrupt work. Vandalism, and pilferage by neighborhood adolescents, for example, can result if teen-agers feel they have their parents' tacit approval. Owner-builders might make the effort to find out what kind of neighbors surround the land they plan to acquire and, if possible, determine how the neighbors feel about an extended construction project. People are always intensely curious about what kind of homeowners are going to live close by, and some friendly overtures can turn otherwise hostile neighbors into helpful, cooperative ones.

This cooperation may be even more important if the builder intends to move a trailer onto the property to live in while building the house. Many owner-builders find this a convenient and economical way to keep close to the building site and to utilize their time more efficiently. This is particularly true during summer months when construction is usually most intense. Small vacation trailers can be borrowed, rented, or even purchased used, then sold when no longer needed. If the building site is any distance from one's living quarters, a trailer can save hours of travel time every day. In some cases, living on the site in a trailer is the only way to prevent theft of building supplies. This is why so many contractors keep a workman living in a trailer on commercial building projects. Some owner-builders simply camp on their land during good weather. One Oregon couple lived in a tent on their wooded building site until they were driven into a trailer by approaching winter. "Our recreation," the woman recalled, "was going into town to take a shower and do the laundry."

Whether building codes allow one to live for an extended period in a trailer or tent may be a consideration in the selection of a site. Again, neighbors can make all the difference, because rarely is anything done about such temporary code violations unless someone complains. Indeed, how strictly building codes are enforced may be of major concern to owner-builders in selecting land. I discuss codes in Chapter 10, but for now it should be noted that communities differ widely in how they administer and enforce building codes. If difficulties are anticipated with codes, another owner-builder in the community is usually the best source of information about the ins and outs of code enforcement. If codes come in conflict with building plans, it is sometimes wiser to seek land in another community.

Besides the human ones, there are physical variables that may affect a self-help builder's selection of a site. Perhaps the most critical, as I have already suggested, is the orientation of the land to the sun, for solar heating must begin with the availability of sunlight. For example, any land that is on the north side of a forested hill is going to present problems in collecting winter sunlight. Obstructions of every kind, from trees to neighboring buildings, can similarly prevent the use of the sun as an energy source. The owner-builder will probably be faced with solar decisions more than a home buyer because his or her house is much more likely to be a process rather than a completed object. Because owner-builders act autonomously, they are able to construct possibilities and the future rather than deal with completions and the present. If solar energy seems too experimental and costly at the moment, very well, build without it, but pay heed to the sun angles, watch rooflines, turn windows south, and sit back and wait. Solar systems can be added later when the technology is ripe. But first be sure the sun is in your pocket.

There may also be soil considerations to be dealt with. In some parts of the country—the Southwest in particular—building with earth is not only traditional but superior for that climate. Many owner-builders in this region make their own adobe blocks and use earth from the building site as mortar and plaster. For such construction projects, soil analysis may be prudent before acquiring land. Similarly, the availability of rock on or near the building site may be critical for those contemplating doing their own rock masonry. Anyone considering building with rock gathered on the site should read Helen and Scott Nearing's *Living the Good Life* for inspiration and advice. It is one of the classics of self-help building literature.

However, using lumber from the site presents numerous problems. It may seem like an attractive way of getting cheap building materials, but owner-builders planning to cut and/or mill their own lumber should be aware of the amount of labor and time involved in such an undertaking and of the difficulties of working with green lumber. Building with logs holds an enduring fascination for Americans, however; I discuss this method of building in Chapter 4.

The topography of the land can also figure prominently in an owner-builder's plans, particularly when a decision must be made about a concrete slab, a basement, or a crawl space. A slab is a very tempting option for an owner-builder. To prepare for a slab, all that must be done is: dig the footings; position the forms and reinforcement; lay any electrical conduit or plumbing that will go under the slab; and call in a concrete crew. In one day, wonderfully, there is a level building platform—and the floor of the house. However, there must be a level building site in order to construct a slab; otherwise it is costly and impractical. Land that cannot easily be leveled is unsuitable for a slab.

To get the same level platform and floor by building a basement, an owner-builder must pour footings; construct a masonry perimeter wall eight feet high; then place floor joists and a floor deck on top of this wall. (See the alternative method discussed in Chapter 4.) The difference in time, labor, materials, and cost is considerable. But so is the enclosed, usable space gained— it doubles the square footage of a house. A basement is probably the cheapest utility space an owner can build. Unfortunately, the extra cost of constructing it comes at the beginning of the project when money may be sorely needed to get the house framed. Would that the owner-builder, like the architects in Jonathan Swift's Grand Academy of Lagado, could build from the roof down.

Aside from costs, deciding on whether or not to build a basement is related to the topography of the building site; it makes a lot more sense on a sloping site. On level ground, the builder must excavate a six- to eight-foot hole the size of the house, and then run the risk of a leaking, damp, dark basement. On the other hand, the builder who selects the side of a hill or a sloping lot need only make an eight-foot cut into the hill, and half of the basement will be above ground level. It is easily

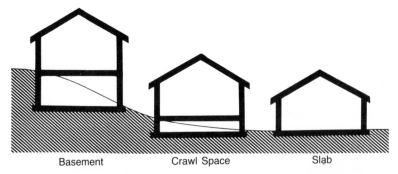

Basement   Crawl Space   Slab

The topography of the land may dictate the optimal house foundation. The sharp slope on the left makes a basement easy to build and keeps the basement dry; the gentle slope in the center is perfect for a crawl space. A slab requires a level building site.

TYPES OF FOUNDATIONS IN OWNER-BUILT HOMES

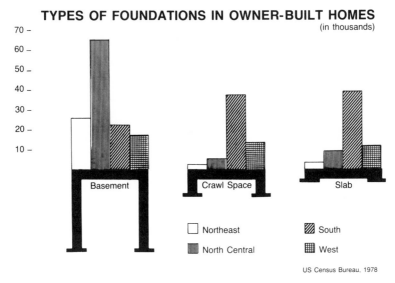

(in thousands)

US Census Bureau, 1978

Whether owner-builders construct a basement, slab, or crawl space will depend in part on where they live. In the Northeast and North Central sections of the country most houses have basements; in the South crawl spaces and slabs are more common. Western owner-builders show no particular preference.

kept dry and there is ample above-ground wall space for doors and windows.

Local conditions—a high water table or a deep frost line—or simply traditional construction practices may dictate a slab or crawl space rather than a basement, but usually the builder is able to choose among the three alternatives.* Each of these foundations is suited to a particular terrain configuration: level ground is good for a slab; a gentle slope is best for a crawl

---

*The potential owner builder will be interested in knowing that the U.S. Census Bureau keeps records of the kinds of foundations owner-builders construct. The rough percentages are: basements 55 percent; slabs 20 percent; crawl spaces 25 percent. These are national averages; there are wide variations geographically. Because of frost problems in the Northeast and North central parts of the country, for example, only a small number of slabs are built. Eighty percent of owner-built houses in these regions have basements. In the South, more than 30 percent of owner-built houses are on slabs; only 25 percent have basements. In the West, about 35 percent of owner-built houses are on slabs and 40 percent have basements.

space; and a steep slope for a basement. Of course, a bulldozer can do anything it wants with the land, and almost any site can be made to accommodate any foundation. The terpitect will find, however, that he or she will be more comfortable with the way the house reclines upon its bed if it is wedded to the landscape and rested gently on the earth.

PINCHING PENNIES:
THE ECONOMICS OF
SELF-HELP BUILDING

One of the first
questions asked of any owner-builder is How much cheaper
was it? The unspoken assumption is that the main reason, perhaps the only reason, anyone would have for building a house
oneself is to save money. I have been challenging that assumption. Although many owner-builders begin their enterprise thinking only about the money they will save, few of them
end this way. My sample of owner-builders has convinced me
that personal satisfaction, self-esteem, pride, assertiveness, and
creativity all ultimately emerge from the experience.

Nevertheless, every owner-builder figures in his or her motivational equation a generous monetary plus. This plus is realized in two ways: by acting as one's own general contractor,
the owner-builder saves the roughly 15 percent profit that a
contractor makes on a custom-built home. (Compared to a tract-
built home, however, this figure may be less than 15 percent;
large developers sometimes operate on a smaller profit margin.)
The other source of saving comes from "sweat equity"—the
self-help builder's own labor. How much this amounts to ob-

viously depends on how many of the various subsystems of the house are self-built. A small minority do all of the work themselves, the majority do some—perhaps half—of it, and another minority subcontract all of the house systems.

Because no two owner-builders put exactly the same amount of their own or their family's labor into a house, it is difficult to arrive at an average amount saved by self-help building, something every cost-conscious potential owner-builder is interested in knowing. Labor costs vary throughout the nation, and this influences the dollar value of the owner-builder's sweat equity. In areas where labor costs are high, sweat equity savings will be substantial. The size of the house can also affect savings. A larger house may save the builder more than a smaller one, since the cost per square foot of a house is not distributed evenly throughout the entire floor plan. The kitchen and bathrooms are obviously the most expensive rooms to build, and every house has at least one of each. Enlarging a house by including more relatively inexpensive space, such as that found in bedrooms, will increase the amount saved by owner-building. A 1500-square-foot, two-bedroom house, for example, would save less than a 2,000-square-foot, four-bedroom house with two baths. The additional 500 square feet of living space in the four-bedroom house is built with the least expensive materials that go into a house—framing, siding, roofing, and interior finishing—and labor for these systems is the kind owner-builders are most likely to do themselves. Since savings depend on a comparison between costs and the market value of the finished house, the larger house may also command a greater market price, which amounts to a great saving to the builder.

Location of the house can also influence its market value. A house built in a rural community may have a market value considerably less than the same dwelling built in the suburbs of a metropolitan area. Therefore, the savings on the suburban house would be higher, even if the cost of materials and the amount of labor in both houses were identical. For example, let us assume that materials for a house cost $40,000 and the builder did half of the labor. The completed rural house might well be valued at $50,000, whereas the suburban house, in a more desirable neighborhood, might have a market value of

$60,000. For the rural house, the saving would be 20 percent; for the suburban house it would be 33 percent.

I asked my sample of owner-builders how much they saved by self-help building, and I analyzed these responses closely. My conclusion is that most owner-builders inflate the amount of savings realized on their homes, and they do it for several reasons. Partly it is no doubt wishful thinking—the desire to feel that one has realized the greatest possible financial reward for one's building efforts. But I also feel that most owner-builders' houses cost much more than they are aware of—partly because of poor record keeping and partly because of unrecognized, hidden costs. One frequent lament of owner-builders is, "I wish I had kept more accurate records." Many home builders, myself included, do not take the trouble to set up a balance sheet for *all* building costs and keep to it rigorously. Many purchases, particularly small ones, are simply not recorded; the cost of necessary tools is not included; and other costs are overlooked. How many owner-builders, for example, include transportation costs—the hundreds of times they use their car or truck to get to the construction site or run down to a building supply center for materials? What about hidden labor costs—feeding or entertaining the "free" help that stops by on a weekend? What about the loss of income from a summer job the owner-builder does not take in order to have the time for construction, or the "moonlighting" jobs that are abandoned during the construction process? What is the monetary value of the distraction in the builder's business, profession, or job caused by the time and energy given up to the house? These are all a legitimate part of the owner-builder's costs and are rarely recorded.

At the other end of the equation, owner-builders are frequently unrealistic about the worth of their finished houses. Since few builders immediately sell their houses, the value placed on them is usually a guess based on what similar houses are selling for in that neighborhood, or on a cost-per-square-foot formula. As every real estate agent knows, owners' notions of what a house will fetch on the market are often overly optimistic. Owner-builders who have first had a contractor bid on their house plans are in a much better position to judge how much is saved, but even this can be inaccurate because of the numerous changes and additions that usually occur during construction. The fact that a house may take several years to complete also complicates any attempt to measure savings. The market value of the house increases yearly according to the annual rate of housing inflation; at what point is this market value to be measured?

What then can be said about how much an owner-builder saves? I have heard from owner-builders who kept meticulous records of all of their costs and who then had their houses professionally appraised upon completion. Those who did much of the work themselves and used new materials entirely, realized savings of roughly one-third of the cost of having the house professionally built. This is about what I calculate I saved on my own house, and I did all of the labor except grading, septic tank, basement floor, and carpeting. But then my house has quite a few "luxury" items in it; I could probably have saved another 5 to 10 percent had I been interested in building the most inexpensive house possible and if I had been more skilled. Quite a few of the owner-builders I queried claimed to have built their houses for half of what the contractor price of the house would have been, by doing just about all of the labor themselves. Some of these houses were built up to ten years ago, however, and it is doubtful whether these same savings could be realized today unless used or salvaged materials were used. In recent years, the rate of inflation has hit material costs harder than labor costs. This means that the owner-builder will pay more for materials and will save less on sweat equity.

PAY-AS-YOU-BUILD CONSTRUCTION

The best piece of financial advice that could be given to anyone contemplating building his or her own house is: if at all humanly possible, *don't borrow money*. Pay for the house as you build. I know that for those who have watched the median price of a house in this country leap upward at the rate of thousands of dollars every few months, this may sound like a cynical joke. It certainly cannot be accomplished by everyone, but Census

Bureau figures show that more than one-third of all owner-builders manage to finish their homes without borrowing money. There is no way of knowing how many of these are first-time builders, but it is my belief that the number is not large. There is no reason to conclude that owners who are building for the first time are any more numerous than home buyers who are entering the housing market for the first time. Both have been priced out of owning their homes by the dizzying spiral of housing costs. The majority of home buyers in this country in recent years have been homeowners who are "buying upward," selling their houses and purchasing larger, more expensive homes in better neighborhoods. As with home purchasers, the majority of owner-builders who construct their houses for cash are probably on their second or third house, and a surprising number are in their middle or later years. Building one's own home is not necessarily for the young, as I demonstrate in later chapters. The young, however, are the ones who must face the seeming impossibility of accumulating enough money to reap the rewards of building for cash.

First, consider exactly what the rewards are. Let us suppose that three families take out loans to build new houses—$50,000 at 15 percent for twenty-five years. The Greens have a combined income of $20,000; the Greys, $40,000; the Blues, $100,000. (The Greens probably could not qualify for a loan, but for the sake of argument, let us suppose that they did.) Each family would face monthly interest-principal payments of $640. At the end of the first year, each would have paid $7,680 to the loan institution, of which $7,480 would have gone for interest and $200 for principal. Each family would still owe $49,800 of their $50,000 loan.

However, when they paid their income tax and declared the interest paid on their mortgages as deductions, a different picture emerges. After considering the amount of deduction allowed, you could say the Greens' payment was $460 a month; the Greys', $370; and the Blues', $290. The tax laws favor the affluent house buyer at the expense of moderate- and low-income families.

At the end of seven years, the average time Americans stay in a single-family home before selling it, each of the families

Effect of Interest Rate on Cost of a $50,000 Loan Over a Twenty-Five-Year Period

Interest Rate* (in percentage)	Monthly Payment (principal & interest)	Total Interest (over 25 years)
8	385	65,500
9	419	75,700
10	454	86,200
11	490	97,000
12	526	107,800
13	563	118,900
14	601	130,300
15	640	142,000

*Soaring interest rates on mortgages make cash building increasingly more attractive. The amount an owner-builder pays in interest, before income tax deductions, has doubled in the past five years.

would have paid $53,760 to the loan company, of which $2,270 would have gone toward reducing the principal. Each family would still owe $47,730 on their $50,000 loan. After tax deductions, however, the Greens would have paid roughly $38,500; the Greys, $31,000; and the Blues, $24,250.

Given a credit system that is designed primarily for the profit of banks and real estate interests, it behooves owner-builders to utilize every resource to avoid borrowing money. Builders must come to realize that their friendly banker, handy building and loan association, and helpful real estate agent are the enemy; they will keep you from owning what you build, and ownership is one of the continuing pleasures of a self-built home.

When confronted with decisions about how to pay for their building enterprise, owner-builders should keep in mind that by borrowing money, they are trading interest payments, not for ownership, but for immediate possession. For those willing to bide their time, to forgo immediate possession of a home, it is within their grasp to escape the banker's interest trap, to possess *and* own their house.

Obviously, a pay-as-you-go home, or what I will call a cash-built house, must begin with some capital. Ownership of a piece of land, a fulltime job, and no credit payments are a start. The absence of credit obligations is important, for whether it is paying off a car, paying for bank card purchases, or paying off furniture or appliances, it means that your income is encumbered. It is essential that each paycheck is free if a house is to be paid for in cash. If a new car has been purchased, it might be sold or exchanged for a used pickup, an invaluable transportation tool for the owner-builder. Since building as you go involves numerous small purchases of construction materials, a pickup or station wagon is highly desirable.

While it is advisable to be completely free of credit payments, it is equally useful to have a good credit rating. Cash building will inevitably involve some short-term borrowing, and this is both acceptable and necessary. Credit cards from a bank, building supply center, or department store can enable the cash builder to continue working on the house when he or she would otherwise run out of materials before the next paycheck. An extra load of framing lumber may be needed before the cash is available, and a credit card can keep the builder in business. It is absolutely essential, however, that all short-term borrowing be immediately paid off, and not allowed to become a permanent credit obligation that must be met with payments from income.

A cash builder will have to plan construction strategies carefully. The major goal is to get a roof up as quickly as possible, to get "dried in." Once the house frame is protected from the weather by a finished roof, a financial breather can be taken and money accumulated. Try to plan this breather for the winter months since they are poor for building. It is this initial spurt of building, then, that places the greatest pressure on the cash builder's financial and human resources. If he or she can get dried in and still be afloat financially there is an excellent chance of finishing the house without borrowing.

Every owner-builder should think of the house as a series of discrete building systems, each a complete construction unit. Four of these systems are involved in getting a house dried in: grading, foundation, framing, and roofing. The first two are not affected by the weather once they are completed, and can await the next two systems if the builder wishes to accumulate more money. Grading usually involves clearing the land of brush or trees, running a driveway to the building site, excavating for a basement/crawl space, or leveling for a slab, and digging trenches for footings. Although a bulldozer is obviously needed for heavy earth moving, cash builders can do much of the clearing themselves with a rented chain saw and a pickup truck. And they can dig their own footings, as I did, with a pick and shovel. The physical exercise is good conditioning for the work ahead.

Setting up batter boards and constructing forms for the footings is work that all owner-builders should do; it is an act of centering that puts them in touch with the earth and establishes with sensory immediacy the abstract geometrical principles that will guide their enterprise—level and square. Pouring concrete footings is simple enough, but pouring and finishing a slab calls for an experienced crew if it is to be satisfactorily level and smooth.

The cost-conscious cash builder should seriously consider

In order to finish a house without a mortgage, owner-builders usually have to do much of the work themselves. Many amateurs learn masonry skills by laying concrete block footings and then work up to brick and rock masonry.

doing the masonry work for either a crawl space or a basement. (I discuss the skills involved in masonry in the next chapter.) Laying concrete blocks for a basement or laying bricks or blocks for a crawl space is a convenient opportunity to learn and practice masonry skills. Aesthetic errors are easily concealed here, and you will be more ready to tackle other masonry projects.

Framing is one of the costliest of all house systems because of the amount of material it takes. The frame is the skeleton of the house upon which all other systems are hung. It is usually built entirely of structural lumber, 2 × 4's, 2 × 6's, 2 × 8's, and plywood. By building according to the OVE system, however, the cash builder can limit these materials to the absolute minimum compatible with sound construction practices and safety. The OVE system will also reduce the amount of labor needed to frame the house and enable the cash builder to get dried in more quickly. If roof trusses are used, only the exterior walls and roof need be erected; interior walls can wait until later.

I would strongly urge all owner-builders to frame their own houses; the cash builder will find it absolutely necessary if he or she is to stretch limited capital. My survey of owner-builders shows that of all house systems, framing was one of the easiest and most rewarding. Aside from a few tricky details, a "stick house" frame is completely standardized and fully outlined in literally hundreds of publications. And if there are still any questions about it, the owner-builder can walk to virtually any construction site in the community and see a standing example of house framing. The fear that it can be constructed incorrectly and that somehow the structure may collapse like a house of cards is wholly unfounded. If standard building practices are followed, the only thing that can go wrong is that the house may not be perfectly level and square. If this does happen, the owner-builder will have the consolation of knowing that his or her house is like almost every other professionally built house in the community. Few houses are geometrically true. That is what molding is for—to cover up the gaps.

Moreover, framing one's own house produces more instant pleasure than any other house system, because everything happens so quickly. Where yesterday there was only an empty floor, today there is a wall; inside has been separated from outside. True, the "wall" is only studs, and the windows and doors are open air, but the terpitect's imagination gives to airy nothing a local habitation and a name. I've had a number of owner-builders report that the most satisfying experience of the entire project was when the roof went up; for many, a roof remains the universal symbol for shelter.

To make the house weatherproof once the frame is completed, all you need to do is cover the stud walls with fiberboard sheathing or building paper if sheathing is omitted. A better choice is one-inch-thick sheets of rigid polystyrene plastic insulation. Asphalt shingles or wooden shakes complete the roof. Again, cash builders will want to place the roofing on themselves; it is work that requires little skill. If money is especially scarce, roofing can be delayed by covering the plywood decking on the roof with building paper and nailing it firmly to prevent wind damage. I did this on my house; I had to repair rips in the roofing paper periodically, but it kept my house dry.

Wall openings for windows and doors and the roof opening for a chimney can now be covered with polyethylene or plywood until the cash builder is ready to continue the project. While more cash is accumulated, there is no need to remain inactive, however. Many "labor only" tasks connected with later systems can be performed: drilling holes and cutting openings in framing members for electricity and plumbing, cutting stair stringers, excavating outside plumbing ditches or retaining walls. I was a cash builder, and although my money ran out on numerous occasions, I never once was forced to stop working for lack of something to do.

As the cash builder hangs more skins on the house frame— siding, windows, doors, stairs—the time will get closer when even more financial leverage can be gained by moving into the unfinished house. Most cash builders move into their houses as soon as kitchen and bathroom facilities are in, unless occupancy codes forbid it (and even then, some manage it). In my survey of owner-builders, I found that roughly half completed their houses before moving in; the other half moved in during widely varying stages of completion. Some literally

Many owner-builders economize on building by moving into an unfinished house. Living in a house under construction can be quite comfortable.

"camped in," with no electricity or plumbing, but most did not move in until the minimal comforts of a kitchen and bath were finished. Most had some of the interior walls still unfinished. "We hung bedspreads over the studs in the bathroom," one owner-builder reported, "but the family got used to it quickly enough." The obvious financial advantage of moving into an unfinished house is that the rent or house payments from the previous dwelling can now be spent on construction. I discuss the disadvantages in Chapter 8.

From this short survey of how a pay-as-you-build house might be managed during its early, most difficult, period, the reader can get some idea of the human resources necessary for such an undertaking. The cash builder must have a strong desire—better, a need—to finish a house without going into debt, supported by what might best be described as the Puritan work ethic: thriftiness, discipline, patience, self-denial, and tough-mindedness. These qualities, which may be split between a man/woman team, will permit cash builders to stretch the building program over an extended period of time, to forgo present comforts for future rewards, and to subordinate every other consideration to the uncompromising goal of owning a house outright.

BORROWING MONEY

Not everyone, alas, can be a cash builder. One's living expenses may be too high and one's income too low; there may be land to buy and money for nothing else; there may not be time; there may not be patience; there may be too little self-discipline. And so the only way to build is to borrow. Before going to a bank or building and loan office, however, every owner-builder, especially young couples who are planning their first home, should try borrowing from their families. I found that a large number of those in my survey of self-help builders were successful in getting a family loan. If family members can be convinced of the economic advantages and the potentials for personal growth in self-help building, they may be persuaded to lend, at no interest, the seed money needed to get a house framed and dried in. Such a loan would make a perfect wedding gift and could be repaid after the house is completed.

Those who are compelled to borrow from a bank or loan association will usually find themselves in an unenviable position. To see why, look at an owner-builder applicant from the point of view of loan institution officers. They are accustomed to doing business through well-established real estate channels. When they lend money to a real estate developer or contractor, they are reasonably sure that the houses to be constructed will be completed and salable; there will be collateral for the loan. But what assurance do they have that the amateur house builder, who has probably never before built anything bigger than a breadbox, will ever finish a house? And if it is finished, will it be worth the money lent on it? Where is the collateral? It follows from these questions that a loan institution will be much more inclined to lend money to an owner-builder who has some history of construction experience, a record of finishing some kind of structure, whether it be a barn, a garage, or a house.

Census Bureau statistics show that of the roughly two-thirds of all owner-builders who finance their homes, all but 5 percent get their money from conventional mortgages. The obstacles thrown up by lenders often cause lengthy delays. One owner-builder reported that his construction was postponed a year while he attempted to get a loan. "Nobody believes you can build unless you are a contractor," one builder complained. "To get a loan you need sheer persistence and a gut-burning desire to build your house!"

A number of owner-builders agreed that banks and loan institutions insist upon the security of working with a contractor and will allow the owner to be his or her own contractor only with great reluctance. "They were skeptical," said one builder. "I needed a lot of preplanning to convince them." Some lenders put a time limit on finishing the house, and draws on construction loans can usually be made only after the work is completed. This sometimes presents problems in purchasing materials. Some self-help builders circumvent this by purchasing materials on credit and then paying for the purchases when the draw on the construction loan is made. At the completion of the house, the construction loan is invariably converted to a long-term conventional mortgage.

On the other hand, a large number of my respondents reported that they had no difficulty at all borrowing money. The state of the national economy, as reflected in the availability of funds for lending by building and loan institutions, has a lot to do with the willingness of an institution to finance an owner-built house. When funds are scarce, the available money will go to the most secure risks. Many banks or lending institutions are willing to accept a deed to the land the house is to be built on as collateral for the loan. The more of his or her own money, whether it is in land or cash, that an owner-builder brings to the building project, the easier it will be to borrow. A bank will be much more willing to lend half the cost than 90 percent of the cost of an owner-built house. Also, the borrower with little cash on hand may find that a higher interest rate must be paid for a loan.

If the owner-builder's land is not entirely paid for and there is only a small cash reserve, borrowing may prove to be very difficult. A good number of owner-builders acquire their property through a land contract—a small down payment followed by monthly installments. If financing is obtained on the house, the unpaid land contract can often be added to the long-term mortgage. Obviously, the self-help builder must be certain that any land contract provides the right to start building before the land is paid for, and the right to prepay the balance without penalty.

FINANCING PACKAGE HOMES

One advantage of buying precut homes is that the manufacturer will sometimes aid the builder in financing the house. Insofar as the package home could be considered a disassembled product, a loan on one potentially represents less of a risk to a lending institution than a construction loan. If the buyer were to default on the loan, the bank could have a contractor finish the home from the precut materials on hand. It might, therefore, be easier for an untried owner-builder to get a loan on a precut home.

A few packagers finance the precut materials themselves. An approved builder, even one still paying for the land, can construct a house with virtually no ready cash. The builder pays monthly interest on the cost of the materials until the house is complete; he or she then gets a conventional mortgage and pays the packager for the materials.

This approach runs counter to the conviction by many packagers that lending money to self-help builders is too risky. Many "shell home" manufacturers—sellers of homes with unfinished

A package home that was completed by its owners. Many package home builders have the shell constructed by professionals and finish the interior themselves. (Photo: Miles Homes, Bill Hedrich, Hedrich-Blessing)

interiors—went under in the 1960s because they financed homes that were never completed. When precut materials are financed by the packager, customers are carefully screened: one firm accepts only 50 percent of the applicants. The two qualities looked for in a self-help builder are a strong desire to own a home and some construction ability. One packager claims that 70 percent of its purchasers do all of the work themselves and the other 30 percent complete at least half of the house themselves. The purchaser is allowed three years to complete the home; after that time the total cost of the package is due.

This kind of financing plan no doubt allows a number of owner-builders who otherwise could not manage it to build homes. The cost, however, is high. During the three years that the house is under construction, the builder will pay monthly finance charges on the complete cost of the package. These charges can total from $5,000 to over $7,000, depending on the size of the house and current interest rates. If the home is finished more quickly, the amount would be less. The monthly payments are a total loss, for when the house is finished, the builder still owns not a stick of the lumber in it; all of the payments went for interest. The house financed in this manner costs considerably more than one constructed by an owner-builder who obtains a standard bank construction loan on building materials as they are needed.

CUTTING COSTS

Whether owner-builders borrow money or pay cash, they are often in a position to build much more economically than anyone else. One reason why there are houses—thousands of them—that were constructed for a fraction of their completed value is that their builders were able to cut the cost of materials to the bone. Unlike the contractor or housing developer who must buy standardized materials for standardized labor practices, owner-builders are open to choice of materials. Their labor is flexible, and time is not necessarily money as it is to a professional.

Many amateurs build with used materials—dismantled homes, barns, or outbuildings. Heavy timbers and brick from buildings that have outlived their usefulness can be recycled in an owner-built house. Windows, stairs, and paneling of every kind are often obtained for a fraction of their new cost. Adapting these used materials into a modern house system usually takes a lot of time-consuming custom fitting, but where money is scarce the time is willingly spent.

The self-help builder is also frequently able to buy materials on sale, either because items are discontinued, in odd lots, or simply because the supplier is overstocked. I purchased the roofing for my house at a considerable saving because the building supplier had ordered, for a contractor, too large a shipment of a premium shingle he did not ordinarily stock. By judicious shopping, an owner-builder can often realize savings on structural lumber, one of the costliest items in a house. The builder should beware of compromising on lumber grade where engineering considerations are involved—as in girders, joists, or rafters—but where spans are short, less expensive grades of lumber, especially if they are kiln-dried, are sometimes adequate. I have discovered, however, and many owner-builders concur, that the savings realized by using substandard grades of wood often do not compensate for the frustrations of framing with warped or twisted lengths of lumber. I have purchased utility grade 2 × 4's, but they did not insinuate themselves into my permanent framing; instead, they went for such temporary needs as bracing, scaffolding, or concrete forms.

Professional builders usually buy about 5 percent more materials than actually needed, to replace the waste and pilferage that occur in production building. Owner-builders need not do this because they can take the time to utilize materials that are ordinarily wasted by professionals. Owner-builders should remember, however, that even if they overbuy, the extra materials are money in the bank. The cost of building supplies goes nowhere but up, and owner-builders are sure to use the materials later in their building schemes.

Quite a few of the self-help builders in my survey were able to buy their building materials wholesale, usually by establishing themselves as a contractor and getting contractors' prices.

But the majority probably share my experience of paying retail for most of their building materials. I was, however, able to buy all of my kitchen, bathroom, and plumbing supplies at wholesale simply because I established a cordial relationship with the manager of the supply house. Most building supply establishments have a layered price structure, and owner-builders can often get a discount on their purchases simply by asking for it, especially if they are buying in quantity. If self-help builders take the attitude that they are shopping in a Turkish bazaar where no prices are ever fixed, they will find that absolute bottom prices sometimes have a way of going even lower. An owner-builder should also never be intimidated by signs that say Wholesale Only. Wholesale suppliers are where owner-builders want to shop, and sizable purchases will often establish them as wholesale buyers.

Some of the best sources of information on prices come from other owner-builders. Anyone building a house will soon be in communication with other home builders; this, as I discuss in Chapter 8, is one of the great strengths of the owner-builder movement. Every owner-builder I have ever talked to has been delighted to discuss building strategies with me and to offer help if it is needed. Cost information is crucial to every amateur builder, and no owner-builder should ever be reluctant to ask a fellow builder where to purchase supplies economically. This will invariably prove to be one of the most knowledgeable sources of price information available.

THREE

DON'T KNOW NOTHIN':
MASTERING SKILLS

One way to talk about house building is through the language of systems analysis, an intellectual tool that breaks down complexities into a series of simplified relationships. Systems theory, somewhat in the way that grammar is used to analyze language, permits us to see much more clearly, by abstraction, how processes and things work. For example, a systems analyst would define a house under construction as a fairly regular, segmented system consisting of a series of subsystems added at sequential intervals. This translates into what anyone who has ever watched a house being built has observed: workers construct it in a number of separate operations, or subsystems, each independent of the other—the foundation is dug, framing is put up, bricks laid, plumbing and electricity installed, etc.—one after another.

The subsystems, roughly in the order of each operation, are: design; financial management; grading; foundation; framing and roofing; electricity; plumbing; sheathing, or outer skin (including masonry and fireplace); heating/air-conditioning; interior skin (sheet rock and paneling); mill work (doors, windows, trim, kitchen and bathroom cabinets); painting and wall covering;

and, finally, floor covering. Each of these requires a separate set of skills, some of which the owner-builder may already have. Many self-help builders enter their projects with virtually no trades skills, intent upon on-the-job training for all of the house systems.

The owner-builder has a number of options in dealing with skills: he or she can forget about attempting to master a particular skill and hire a professional to complete that particular subsystem (see discussion in Chapter 8); he or she can enroll in a course at a local educational center and acquire skills under the guidance of an instructor; the builder may be able to be apprenticed to a skilled worker on a part-time basis; or the builder can master the skill alone. The skill does not necessarily have to be practiced on one's own house; it may be acquired by helping another owner-builder with his or her project.

In my sample of owner-builders, I found a significant number who acquired carpentry, masonry, plumbing, or electrical skills by taking courses at trade or tech schools. Many communities offer such courses, often through the extension division of a college or university, sometimes through local high schools. A number of school systems throughout the nation now offer courses in house building. Schools devoted entirely to house building have sprung up on the East and West coasts. The Housing Institute and Cornerstones in New England, and The Owner Builder Center in the Bay area of California all teach house building from planning to finishing a house. The largest and most ambitious is The Owner Builder Center. In two years this has grown from a two-man operation into a thriving non-profit educational institution which offers classes and seminars for more than 2,500 students yearly. The phenomenal growth of The Owner Builder Center demonstrates the potential for educational institutions devoted exclusively to house building.

Would-be owner-builders who are apprehensive about launching a building project about which they know virtually nothing might investigate the educational opportunities in the construction trades. If they have the time to take a course in, say, carpentry or masonry, they will gain not only the specific skill studied but also the self-confidence needed to teach themselves other skills. It is not necessary for self-help builders to excel in all building trades; they will find that a number of skills overlap, and that proficiency in one will sometimes supply competency in another.

Most owner-builders come to their projects with some kind of construction experience, if only from doing odd jobs in home repair and maintenance. An amateur builder who has previously constructed a boat, who has done woodwork or refinishing as a hobby, who has built models, or who likes to tear a car apart will find that all of these activities are related to house-building skills. Many owner-builders are blue-collar workers—heavy machinery operators, truck drivers, mechanics—people who are used to working with their hands. Since so much of house building involves arduous physical labor, vocations or hobbies that are labor-intensive are good preparation for self-help building.

Acquiring skills by first constructing a smaller building has a great deal of merit. One of the problems with on-the-job learning is that the novice usually does not acquire the ability to do professional-quality work until the particular system is nearly finished. As a result, early stages of the job often show the amateur's hand. By constructing an outbuilding—a barn, a garage, or a storage structure (which can be used as a tem-

Building a smaller outbuilding, workshop, or study is one way owner-builders often learn skills. This builder wanted to try his hand at a hyperbolic roof design so he built one on his garage.

porary living quarters while the house is being constructed)—
the self-help builder starts his or her house with skills already
acquired. Work will go more quickly and smoothly, and will
look more handsome. The average builder, however, does not
have the time, resources, or inclination to construct a "practice"
building; skills must be learned as the house itself is built.

I have been discussing skills as if they were a single entity;
actually, each house system requires of the builder two distinct
kinds of competence. One is the knowledge of how the particular system is put together. The other is manual dexterity—
the physical ability to handle tools easily and efficiently. Some
house systems are weighted unevenly on the side of knowledge;
others require a disproportionate amount of physical talent.
Electrical work is an example of a building system that needs
very little manual skill—drilling holes and stripping wire are
the most difficult jobs—but a great amount of theoretical knowledge in planning circuits, computing amperage, and connecting
fixtures and appliances. Finishing dry wall, on the other hand,
is virtually all physical skill; the only way to master it is by
repeated practice.

The knowledge of how a building system is fabricated is easily
acquired by reading. There are dozens of manuals available for
every system or subsystem of a house—many of them inexpensive or free pamphlets published by building supply manufacturers. Magazines specializing in do-it-yourself projects also
contain a wealth of miscellaneous information on construction
techniques. Owner-builders should accumulate these as soon
as they start thinking about building a house. The annotated
bibliography in the back of this book offers some reading suggestions, but the literature is so vast, and arrives and disappears
so rapidly, that it is difficult to keep up with it. In shopping
for do-it-yourself handbooks, the self-help builder would be
wise to examine the illustrations closely. Clarity of the illustrations separates the useful from the useless publications. For
instructional purposes line drawings in color are ordinarily superior to photography. Illustrations in such national magazines
as *Popular Science* and *Mechanics Illustrated* are generally
excellent and might serve as models for the kind of visuals to
look for in a self-help publication.

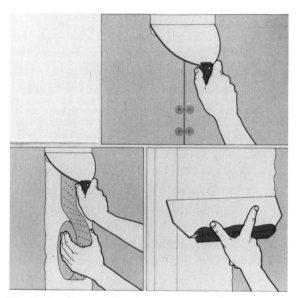

Simple line-drawn illustrations, such as this series on taping sheet rock, are preferable to photographs in construction manuals. (Photo: Courtesy Time-Life Books, Inc., from Home Repairs and Improvement, *New Living Spaces*)

There is a temptation to acquire ever more publications—books, magazines, manuals—and to study the details of house systems that will not be built for months to come. I wasted a great deal of time doing this, mostly because of insecurity about skills that I did not have. If an owner-builder reads about plumbing systems while still working on electricity, much of that reading will have little practical meaning. When the time comes to actually install the plumbing, direct your reading toward particular plumbing problems in specific locations of the house; what had previously been of theoretical interest will now be as pragmatic as a pipewrench.

Even better than reading about how to put a system together is watching it done. Almost every owner-builder I know haunts construction sites. There is no better way for an amateur to learn how a system is built than by watching skilled workers do it. Of course, this is not always feasible, because construction workers are usually at the job during the owner-builder's own working hours, but it is possible to walk through a framed house before the sheet rock is installed and observe the completed framing, roofing, electrical, plumbing, and heating systems.

Just as a teacher who must explain a book to a student reads it differently from a casual reader, a house builder looks at a construction site with specialized vision. He or she sees details a nonbuilder would never notice, and observes with the shock of recognition simple solutions to previously intractable problems. Workers are usually willing to answer construction questions if the inquirer is friendly and deferential. One owner-builder told me that whenever he had a problem he did not know how to deal with, he went down to a local construction site during lunch break and asked one of the workers for information. "They were always friendly and willing to give advice," he said. When I was building my fireplace, I frequently sought the guidance of a wizened, toothless, retired mason who worked part-time at a local building supply firm. He assured me that building a fireplace was an art, not a science, and that every mason did it a little differently. The fact that my fireplace draws like a fine meerschaum is in part due to his willingly offered expertise.

Manual skills are not easily acquired from books; they must be mastered by practice. Still, there are many tricks of the trade that can be picked up by reading. But the best advice for mastering the physical skills needed for house construction is to watch experienced tradesmen at work. The suggestion I made about reading applies here also: look for advice when you are working on a particular skill. If you haven't begun the job, you don't know what to look for. About the time I started to build my basement with concrete blocks, I chanced upon a team of masons putting up a small concrete block building. I watched this team closely for half an hour and learned more in that short time than I had picked up after a month of trial and error. I went back to my basement, and in a few days I had doubled my speed and efficiency. The quality of my work improved, too. I still cannot butter mortar on a brick with the flick of a trowel experienced masons use, but watching enabled me to practice and analyze their movements, and so improve my skill.

Study a craftsman's motions as you would an intricate dance step you were trying to learn. Take notes of the specific body movements. Then practice the movements slowly and methodically—muscles must be taught. Speed and economy of motion come later.

DESIGN SKILLS

The builder who designs his or her own house need not have any extraordinary aptitudes, but should be able to render designs into working drawings. It is certainly not necessary to have a finished set of blueprints for any house system that one builds oneself. It is helpful, but not necessary, to supply a set of blueprints for subcontractors. As long as the self-help builder can produce a scale drawing on graph paper, this will suffice for an experienced subcontractor. However, the self-help builder may need a finished set of construction drawings for the moneylender—who usually wants to see what the plans look like before agreeing to a loan—and for the local building department—which must approve plans before supplying a building permit.

A number of owner-builders in my survey had their designs rendered into architectural drawings by a professional; sometimes an architecture student can be found to do it inexpensively. Many self-help builders find it easier and more economical to purchase a standard set of blueprints from a plan service (see p. 61), and incorporate into these plans any changes they desire. They use these blueprints for the bank and building department. Having a set of blueprints to work from offers a certain amount of security for the first-time builder, especially in the early stages of construction. Once the house is framed, he or she may want to disregard the blueprints and adapt the interior and exterior to an evolving perception of lived-in space.

The owner-builder should be able to read a blueprint; it is a skill easily learned by studying one closely. You will find that you are soon able to visualize three dimensions from a two-dimensional blueprint, but you never really know what a house is going to *feel* like until it surrounds you. The uncertainty

Learning to read blueprints is a skill easily acquired, although it is somewhat difficult to visualize three-dimensional space from a two-dimensional blueprint.

about what a house will look and feel like when finished is what prompts a good number of self-help builders to copy a neighbor's or friend's house. They are not only able to experience in advance the size, scale, comfort, and livability of the house, but they can use it as a full-scale model to supply immediate answers to inevitable construction problems.

THREE-BEDROOM FARMHOUSE   PLAN NO. 7127

This gracious home is well adapted to the family that enjoys outdoor living. The screened porches serve as an extension of the interior spaces while the foyer and attractive fireplace form a focal point for entertaining. Circulation is efficient and located away from the carpeted areas of the house. The work center is located to serve all areas effectively. Closet space is more than adequate. A basement provides additional space for a den, recreation room, or additional storage.

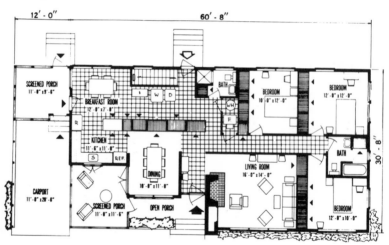

Floor area .... 1645 sq. ft.
Basement .... 430 sq. ft.
Porches .. .. 470 sq. ft.
Carport................ 265 sq. ft.

This three-bedroom house is typical of the plans available free or for a nominal fee through state land grant university agricultural engineering extension offices. (USDA Rural Housing Institute)

INEXPENSIVE HOUSE PLANS

Blueprints purchased from a plan service are expensive, any-where from $75 to $100. Multiple copies cost much more. When you purchase plans from a service, you are buying blind, for the description of a house in a home or plan magazine tells nothing about its construction details—whether it uses roof trusses, whether parts of it are post-and-beam construction, what kinds of materials are called for, how expensive or difficult it will be for an owner-builder to construct.

The U.S. Department of Agriculture has a number of well-designed, single-family house plans available, either free or for a nominal fee. Many of these houses incorporate the results of university research on innovative design and construction.

Plan books listing the blueprints available can be obtained by writing to the Agricultural Engineering Extension Office of your state land grant university. These universities are listed, with addresses, at the end of the Bibliography. Would-be owner-builders can use these plans either as a starting point for designing their own homes or as a way of familiarizing themselves with blueprint reading and standard house construction details.

GRADING SKILLS

Very few men or women who undertake the construction of a house have either the desire or the chutzpa to learn to operate a bulldozer. For most owner-builders, grading is a spectator sport. There are few thrills to rival the sight of a steel-tracked, shovel-nosed behemoth ripping the earth to form a platform for your house. You may have watched a bulldozer in action a hundred times, but it is not the same as watching *your* real estate being pushed with the machine's prodigious power.

The one piece of knowledge that the builder must bring to the grading operation is a firm plan for what is to be done. The house site should be staked, and any trees that are to be saved should be well marked. Bulldozer operators harbor a natural animosity for trees; it is exquisitely easy for the bulldozer to push them down, but damnably annoying to go around them. The operator will try to convince a builder that any tree in his path is unhealthy, worthless, ugly, or dangerous—its roots will destroy the foundation, its sap will waste the family car, and its leaves will clog the drain gutters. The owner-builder must be firm once the decision has been made on which trees to keep, and protect vulnerable ones with temporary fences to keep the bulldozer's blade at bay.

To make sure the site is level, the builder should have a transit, or dumphy level on hand. Find out in advance if the grading contractor will supply one. If not, it can usually be rented, and is a necessity for leveling and squaring the building site quickly and accurately. The transit is a simple instrument to use, and can be mastered with a half-hour's practice. There is a great deal of pleasure for the owner-builder in leveling the site with a transit; it is an information subsystem that controls the massive energy of the bulldozer with a glance through an eyepiece. By sighting through the transit the builder detects that the north side of the excavation is six inches high; with a wave of a hand the bulldozer's bite is set in motion. The

Few owner-builders have the chutzpa to rent a bull-dozer and grade their own construction site. Grading is usually a spectator sport.

machine performs in a trice what aching muscles would take days to execute.

FOUNDATION SKILLS

Footings are often the owner-builder's first introduction to concrete, a material that appears deceptively simple to work with. It is, after all, nothing more than a thick soup that is poured into trenches or forms and then leveled off with a wooden screed. However, avoid mixing it; only if a builder cannot get a concrete mixer truck anywhere near the building site should he or she consider mixing the large batches of concrete that are needed for footings. Ready-mix concrete costs a little more than unmixed, but it saves a lot of time-consuming, back-breaking work.

The main talent needed for concrete work of this kind is a strong back; handling concrete is one of the most arduous jobs an owner-builder is likely to meet. One of the reasons for this is that there is constant pressure to get concrete into the forms before it begins to set, and this means working at a furious pace. And besides, concrete is *very* heavy. One of the worst mistakes a self-help builder can make is to attempt to pour footings without sufficient help, especially if the concrete has to be trucked in wheelbarrows from mixer to footings, as is often the case. A minimum of three, preferably four workers should be available to make this difficult task manageable.

Concrete foundation walls for basements are poured into reusable forms, and for this reason are ordinarily subcontracted—constructing forms for one house is usually prohibitively expensive. Self-help builders who construct their own foundation walls invariably choose concrete blocks for basements, and either blocks or bricks for crawl spaces. (I discuss the skills involved in laying bricks in the section on masonry.) Concrete blocks differ from bricks in two fundamental ways: the work goes much faster, and the blocks are a lot heavier. Most owner-builders who decide to construct a concrete block basement or crawl space foundation will be laying blocks for the first time. This skill can be mastered more quickly if the following strategies, suggested by various owner-builders, are observed:

Before starting, a serious attempt should be made to observe a team of masons working with blocks; watching them will teach more than a thousand pages of instructions.

Purchase a good professional trowel; there is a difference in the feel and handling of a well-balanced tool, and using one is much less tiring than working with a poorly-designed trowel.

Use lightweight blocks instead of concrete blocks. The difference in weight may not seem to warrant the extra cost for lightweight blocks (which are manufactured from lighter aggregate). But it is worth it, especially if twelve-inch blocks are being laid: a twelve-inch lightweight block weighs thirty-seven pounds as opposed to fifty-five pounds for concrete. Masons charge more for laying concrete blocks—for good reasons. By the time you have lifted blocks onto a wheelbarrow, hauled them to the building site, hoisted them up on a scaffold, and then lifted them slowly and gently in place—repeating this operation for hours on end—they get awfully heavy. Masons who have to lift twelve-inch blocks higher than waist level usually work in pairs, each mason grasping one end of the block and swinging it with perfect rhythm into position. I've never

A concrete block basement gives the owner-builder a lot of inexpensive space. The block walls are easily constructed on a sloping building site. Earth will be back-filled against the waterproofed front wall.

talked to an owner-builder who used concrete instead of lightweight blocks and did not regret it.

Pay close attention to the consistency of the mortar. Most of the trouble I had with keeping my mortar joints the same thickness came from using mortar that was either too thin or too thick. If the mortar is the proper consistency, the weight of the block will automatically establish the correct joint thickness. Knowing when it is right obviously comes with experience.

Lay a half-dozen blocks at a time before checking if they are level and plumb. An exception is at the corners. These establish the squareness of the building and should be built carefully. Unless the mortar is too dry, blocks can be adjusted and shifted without harm. The novice tends to check and recheck each block with a level, a time-consuming piece of insecurity.

FRAMING AND ROOFING SKILLS

In my opinion, framing a house provides more sheer fun than any other part of the owner-built house. One reason for this is that minimum labor produces maximum results. A few cuts with the saw, a few dozen nails pounded into boards and lo! there's a section of floor or a piece of wall. Of course, they are only skeletons to be fleshed in after much more time and labor, but for now they enclose space like a picket fence. The skills necessary to produce this instant delight are basic: measuring accurately, cutting squarely, nailing securely. A builder could actually frame a three-bedroom house with four tools: a tape measure, a carpenter's square, a power saw, and a hammer.

If the owner-builder is familiar with standard framing there should be no difficulty in translating blueprints or drawings into a house frame. Taking measurements from a blueprint and executing them on lengths of lumber is as commonsensical as tying a shoelace. This is not to say there will be no mistakes (I discuss those in Chapter 7). The novice would be wise to measure as accurately as possible—to one-sixteenth of an inch—in rough framing. It will quickly be observed that such errors as one-eighth or even one-quarter of an inch cannot possibly be crucial when boards are regularly warped as much as an inch or more; nevertheless, accuracy of measurement is a habit that ought to be cultivated early on, for small errors have a way of multiplying rapidly in repetitive operations.

Equally important is making sure that walls are plumb, and rough openings for windows and doors are square. The builder should make it a habit to try the carpenter's square on every right angle fastened; there are few more sinking feelings than the one which follows trying to place a window unit into a rough opening and finding it doesn't fit because the opening is not square. (I know; I've done it.) Making square cuts is duck soup for any power saw; but if you plan to do a great amount of ripping or make many odd-angle cuts, a table saw or radial-arm saw is preferable. More on these in Chapter 5.

Using a saw and learning to fasten lumber quickly and efficiently with hammer and nails are skills that will develop during framing. The difference between an exasperating experience in cutting and nailing structural lumber and an easy one is often directly related to the number of hands available. Concrete work calls for a brace of strong backs; rough carpentry requires a crew of additional pairs of hands to push, press, force, and hold steady. I did much of the framing on my house myself, and there were many times when I would gladly have traded the self-satisfaction of doing it alone for an extra pair of hands. A man/woman team ought to be able to frame a house comfortably by themselves, although there will be times—raising walls, roof beams, or trusses—when extra help will be needed. I had five friends help raise the four-by-twelve roof beams for my house, and that was none too many. Positioning trusses requires at least three people, more if the span is long and the trusses are heavy. Walls are usually constructed flat on the foundation platform and then lifted into position. This also requires some help, although I did it alone by constructing small sections of wall at a time and joining them after they were raised into position.This requires a lot more time and effort, however, and is another example of where an extra pair of unskilled hands can make an exasperating job simple.

Once the roof is framed and decking is placed over the rafters or trusses, it is ready for shingles. (A flat roof uses built-up layers of roofing paper and hot tar, a job that requires special equipment. Owner-builders who plan on doing the roofing themselves would be wise to design a house with a pitched roof.) Many packages of shingles come with instructions for applying them. Laying shingles is a repetitive job; it requires little more skill than an ability to keep the rows straight. There are two parts of the roofing system that may call for special abilities: working on a steeply pitched roof and installing flashing.

On a steep roof, the builder will have to apply staging, pieces of structural lumber temporarily nailed to the roof and used as a foothold. This is extremely important; more serious accidents occur on roofs than on any other part of the house. Working on a steep roof requires something of the agility of a cat burglar; it is certainly not for an acrophobe, especially if the roof is atop two stories. Flashing, the sheet metal that encloses and waterproofs a masonry chimney, must be cut, creased, pieced together, and fitted into the masonry joints— work that is tough on hands and patience. It is one of those tasks that is not likely to be done again, and takes a lot more time than you probably want to give it; but I found it to be a rewarding exercise in problem solving. You don't really master these kinds of skills the way you master carpentry or masonry; instead, you manage to get the job done simply by putting your ingenuity to work and taking whatever time you need. And if you don't do the clean, professional-looking work a roofer does, it's way up there on the roof where no one can get a close look at it anyway.

ELECTRICAL SKILLS

We surround ourselves at work and home with an infinite array of electrical appliances and devices: We have electric razors to shave our faces, dryers for our hair, washers for our clothes, electric blankets on our beds. The television that we watch for hours on end, the radio, tapes and records we listen to, the heat that warms us, the air-conditioning that cools us—all of our senses are constantly in touch with the products of electricity. It is a modern genie, ready to obey our command at the touch of a fingertip.

Yet, many of us know little more about electricity than medieval man knew about one of its most violent manifestations— lightning. What little we understand is the product of childhood taboos, necessary fears planted in us by parental injunction: "Don't touch the wires; you'll get electrocuted!" It is true that there is virtually nothing else so intimate in our lives that can kill us so quickly (except, perhaps, a jealous spouse), yet few of us are much interested in learning how this friendly, yet potentially deadly, servant works. The guts of an appliance or the electrical system of a house are as mysterious to most Americans as the ganglia of their own nervous systems.

The owner-builder who wires his or her house *must* understand the principles of electricity. Security will come with understanding. Electricity is not a mysterious threat, but simply another house system to be mastered. Electrical theory at the house wiring level is surprisingly uncomplicated; the novice will be able to learn the applications in a single reading of one of the many handbooks on the subject. In effect, applied house wiring is reduced to black connected to black, white to white. The greatest difficulties I had were determining the size wire to use for heavy appliances and wiring a three-way switch.

In my sample of owner-builders, a large number listed electrical work as the easiest of all house systems and quite a few found it the most enjoyable. I agree on both counts. My first acquaintance with electricity when building my own house was not auspicious—I wired my temporary service box with 220 instead of 110 volts and immediately burned out my power saw. ("Hey! Look how fast it's going!") I soon found, however, that nailing up switch boxes, running wire from the service entrance to outlets, and stripping wire were tasks that required no particular skill but were very satisfying. I don't exactly know why; I suspect that it has something to do with the fact that I was doing so comfortably work I had always considered to be "professional," something only an electrician could possibly touch. It was, in other words, the very lack of skill that provided the pleasure. I believe that much of the delight owner-builders get from their projects is of this kind: they find themselves accomplishing with ease a series of tasks that they had previously thought required abilities beyond their reach.

PLUMBING SKILLS

When I was trying to decide whether to install plastic or copper pipes in my house I happened to run into a professional plumber at a local bar one night. I confided in him my indecision and asked his opinion. "Go with copper," he responded firmly. "It's quality." But what about the difficulty of sweating copper joints, something I had never done and had misgivings about? "No sweat," he replied with a self-satisfied grin. "It's a snap. All you have to do is slap some dope on the fittings, heat them up with a torch, and apply the solder until a drop falls off the joint. That's it."

I took his advice and installed copper, a decision I never regretted, but I found that soldering joints was not the snap he described. As is the case with so much of construction work, there is a wide gap between knowing how to do it and doing it professionally. I can honestly say that I never did master the seemingly simple skill of soldering copper pipes together to get the clean joint a plumber gets. It's not only a matter of aesthetics either. When I first tested my water system I found several leaks, and I had to play with the joints for some time before fixing all of them. I never felt entirely certain afterward that any joint I soldered was not going to leak. I found that sweating ½-inch pipe was relatively easy, but ¾-inch pipe gave me trouble because of the difficulty of heating the joints uniformly. Using an amateur's propane torch instead of a professional plumber's torch may also have had something to do with it.

On the other hand, unless building codes absolutely forbid it, waste disposal systems should be fabricated of plastic. Pipes and fittings of such plastics as ABS or PVC have changed the technology of plumbing so radically that it is possible for any self-help builder to install his or her own waste disposal system handily. Not easily—handily. I found that working with these materials is not the child's play that their manufacturers would

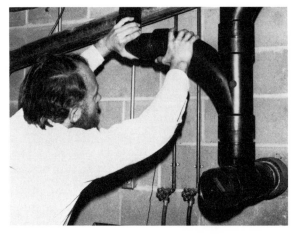

Plastic pipe has revolutionized plumbing for owner-builders. Pipe fittings, which are connected by applying plastic cement and twisting them into position, require little skill to install.

have you believe. It is true that the pipe can be cut with a handsaw (I used my radial-arm saw) and joined by coating cement on male and female joints and forcing them together with a twist. But the virtues of the material are also its faults. Fittings can be joined easily and quickly, but the cement that joins them sets in a matter of seconds and there is no room for error.

In joining large three- or four-inch diameter pipe I found the cement sometimes set before I was able to force the pipes completely together. If the fit of a threaded steel pipe is off a couple of degrees you can give the pipe a turn with your wrench and set it right. If a sweated copper joint is off you can always put a torch to it to loosen the solder and give the joint a slight turn. With plastic, once it is set, it is set for good. You have to be more careful in planning, therefore, than with other plumbing materials. When, inevitably, you make a mistake, you may find that it is not always easy or even possible to saw off the offending fitting and replace it. The one consolation is that the plastic pipes and fittings are relatively inexpensive; if you botch some fittings the loss is not too great—especially if you compare it with the cost of hiring a plumber.

SIDING AND MASONRY SKILLS

Siding, the outside skin that covers the house frame, comes in a wide variety of materials: aluminum, vinyl, plywood, cedar shingles, wood (tongue-and-groove or lapboard), and brick. The builder who has framed his or her house will by this time feel proficient enough with hammer and saw to install the siding without help. Except for brick, siding requires the same carpentry skills that were practiced in framing the house. There is, however, one important difference: siding is finish work. Rough carpentry is always covered with other layers of material, so the main consideration is structural strength. Siding is a final layer, so care must be taken with measurements, fit, and nailing. It is true that poorly fitted, crooked, or badly nailed siding will not affect the structural integrity of the building, and that paint will cover a lot of imperfections. But aesthetic considerations are even more important for the owner-builder than they are for the professional. The amateur's house will be judged by the finish work, and friends and family will apply the same standards in forming opinions about the quality of the work that they apply to professionals. It will not matter that the house is painstakingly framed or that the insulation is as snug as a bear's coat; a couple of visible hammer dents or bent nails in the siding will convince a critical friend that it is schlock work.

Aesthetic considerations are crucial for the owner-builder. The finished house is not a product that was selected and therefore simply a measure of taste; it is an artifact that has been created and is part of what the builder is. It is not only the opinion of others that concerns owner-builders, but also their own self-esteem. Studs Terkel, in his interviews with workers about their jobs in *Working* (New York: Pantheon Books, 1972), cites a stone mason who enjoys going back to look at masonry walls he has completed, but the first thing that catches his eye is the single stone or brick that he feels he had placed improperly. So it is with those who build their own homes; they know every mistake they have made and invariably wish they had corrected each one—particularly when the mistakes are located where they can be seen daily.

Brick veneer is one of the most popular forms of house siding because in many parts of the country brick is equated with quality construction and also because it is maintenance free. In my survey of owner-builders, I found that few were willing to do brick masonry themselves. There are a number of reasons for this, but certainly one of the more obvious ones is that owner-builders think they are incapable of acquiring the skill to lay bricks. Unlike concrete blocks, which are usually either covered by soil or relegated to the basement where they are not very visible, brick veneer is the Sunday dress of a house; any errors or careless work is there for all to see.

Learning to lay bricks is well worth the time and effort because, unlike other house-building skills, it is one that can be put to use later: building walks, steps, or retaining walls. Brick or rock masonry is also a form of construction that is used with ever-increasing frequency in passive solar houses as a thermal heat mass. Those who are designing such a house will find masonry skills invaluable. With the aid of a good handbook, some on-site observation of masons at work, and enough time to go slowly while learning the physical movements, owner-builders can become quite competent at bricklaying. They will never be able to work with the speed and grace of a master mason—the muscles must be trained over a long period of time for this—but they will be good enough and fast enough to get the job done respectably well.

Building a masonry fireplace and chimney calls for additional knowledge, but the masonry skills are no different from those used in a brick veneer wall. There are two ways of constructing a masonry fireplace: you can build the entire system of firebrick, brick, or rock, or you can install a manufactured steel unit and use masonry for the facing and chimney. (It is also possible, of course, to install a manufactured unit with no masonry at all.) A brick or stone fireplace is the most economical to construct, once masonry skills have been acquired; but heat-circulating, manufactured units are much more energy-efficient and can be used as a supplementary source of heat. In a traditional masonry fireplace, most of the heat goes up the chimney, and there can be a heat loss when the fireplace draws heated air from the room up the chimney.

Rock masonry is one of the most creative of all owner-builder crafts. The pleasures of fitting together the stone mosaic of a fireplace is matched by the pleasure of the finished hearth.

Fine rock masonry work such as this does not come easily for an owner-builder. A wall that "reads" as beautifully as this one takes years of experience to produce. This is why so many builders find working with rock one of the most challenging and creative crafts of all.

Getting the correct dimensions of the firebox is crucial if the fireplace is to draw properly and reflect its heat out into the room. These dimensions were once considered a trade secret of master masons, but they are now published in tables in many construction handbooks and should be followed closely. Constructing the firebox of firebrick and fire clay (not regular mortar, which cracks with heat) is another of those tasks that are not likely to be done more than once and which take vast quantities of time and patience to get right. The satisfaction of finishing a masonry hearth, however, justifies the time. Mine took months to complete, but it is the emotional core of my house; the time and effort continue to pay rich dividends in pleasure.

Rock masonry appeals to many self-help builders because it enables them to use a material that is often free for the hauling. Some of the hardest work in rock masonry is finding rocks, hauling them to the building site, cleaning them, and chipping them to size. In spite of the physical labor, there is an aesthetic involved in rock masonry that is missing from every other form of building; and many amateur masons discover to their surprise that constructing a rock wall is addictive. There is probably as much satisfaction derived from completing a rock veneer or rubble rock wall as from any other construction task. Fitting rocks into the growing mosaic of a wall and working out a pleasing pattern of shapes are essentially artistic concerns. Their mastery does not come easily, but the pleasures derived from them are intense.

HEATING AND AIR-CONDITIONING SKILLS

Because heating and air-conditioning are high-technology systems, they present problems for the owner-builder. Installing heating and A/C, unlike nearly every other house system, may involve skills that are beyond the capacity of the amateur. However, my survey of owner-builders showed a surprisingly large number who installed their own heating systems and a smaller number who also installed air-conditioning. I don't believe this represents a national average; my sample includes a disproportionately large number of builders who did all or nearly all of the work themselves. Nevertheless, a great many more self-help builders tackle heating and air-conditioning than one would think. Census Bureau statistics give some indication of why this is so.

The Census Bureau records the kinds of heating systems owner-builders install and the kind of fuel they use. The percentage of self-built houses using gas is less than the national average of contractor and tract homes; the percentages for electricity and oil are above average. Owner-built houses also have an average of 10 percent more baseboard electrical systems and 10 percent more room or space heaters than purchased homes. These statistics indicate that self-help builders are much more likely to choose a heating system they are able to install themselves with relative ease. Baseboard electrical heating, for example, is no more difficult to install than a light fixture. I installed a supplementary unit in my bedroom in one day. The same is true of room or space heaters and floor or wall furnaces, whether they are fueled by electricity, natural gas, or propane.

Central forced-air systems and hot water or steam systems, on the other hand, are sheet metal or plumbing jobs that require a great deal more skill. The systems must also be designed: heat losses must be calculated, the size of the furnace determined, and duct or pipe runs figured out. This is not as complicated as it sounds, for houses of a given size are usually fitted with a heating system of about the same capacity. It has been the practice in the past to install heating units that are oversized (they cost more and the contractor makes more profit), but these are inefficient in the use of fuel, and self-help builders who are installing their own systems can put in a unit that is no larger than needed. In a well-insulated house, the heating system can be appropriately reduced in size.

I installed a forced-air heating and air-conditioning system, and perhaps my experience will give some indication of the skills necessary. I purchased a used electric heater and air-conditioner from a friend who was switching over to a heat pump. I approached the design of the ductwork with some trepidation (I knew absolutely nothing about it when I began), but after reading up on how a forced-air unit operates I designed a system that not only works perfectly, but has incorporated into it a flexibility unknown in most contractor-installed systems. Having installed a high-low return with dampers on all ducts, I can tune the system to whatever climate conditions prevail and to my family's individual living style.

Sections of duct can be purchased in an infinite variety of sizes and this minimizes custom cutting and fitting. Aside from the plenum, the distribution unit that sits atop the furnace, which was built for me at a sheet metal shop, I bought all of the components for the system off the shelf. I must admit that fitting large sections of ductwork without help was one of the more frustrating tasks in the entire housebuilding project. Sheet metal is an intractable beast with a mind of its own and edges like teeth. It takes two pairs of gloved hands to tame it. Wiring the heater and air-conditioner was no different from wiring the electric oven and range, and hooking the air-conditioner to the system was a standard plumbing job—soldering joints. When the system was completed I called up a local air-conditioning serviceman, who charged the air-conditioner with Freon, checked the system for leaks, and gave it a clean bill of health. As with every other system that owner-builders design and complete themselves, the sense of pride and self-satisfaction I felt when hot and cold air flowed through my house was intense, all the more so because I had never dreamed I was capable of doing something as "professional" as installing a heating system.

Latest Census Bureau figures show that more than one-fourth of all central heating systems installed in owner-built homes are now heat pumps, and that figure will no doubt continue to increase. Although a heat pump is the most desirable kind of electrical heating in many parts of the country because its heating efficiency is superior to a forced-air system, manufacturers have not encouraged do-it-yourself installation, arguing that the technology is too complex for the amateur. By voiding the warranty unless the system is installed by a licensed contractor, the manufacturer can severely limit owner installation. Sears, that bastion of self-build merchandising, once declared that its heat pumps were "not for do-it-yourself installation," but it has recently dropped that line from its catalog. Sears includes with its split system heat pump an owner's manual with "complete instructions" for installation.

The history of heat pumps is similar to that of most high-technology home equipment; with time the systems have become increasingly reliable and durable, and installation has been standardized. By discussing installation with the right dealer, owner-builders might be encouraged to go ahead on their own, with the dealer supplying back-up help if needed. A house builder who wants to realize the considerable savings involved in self-installation of a heat pump and who enjoys the challenge of working with high-technology systems should shop around and look over a heat pump installation before making a decision. A certain amount of daring is involved in a positive choice, but not as much, I suspect, as heating contractors would have us believe.

One other increasingly popular kind of heater has found its way into owner-built homes recently—the wood stove. Because of the escalating costs of central heat, thousands of owner-builders are designing wood stoves into their homes, either to supplement other heat sources or to heat an entire house. The controversies among wood stove advocates over the best kind of stove—cast iron or steel, lined or unlined, baffled or unbaffled, updraft or downdraft—are, excuse the pun, heated. Builders who are considering installing a wood stove in a new house or retrofitting one into an existing house should talk to as many stove owners as possible before making a decision. Auburn University has established a testing laboratory for wood stoves and, with the cooperation of the Fireplace Institute, a trade association, is establishing energy efficiency ratings for stoves and manufactured fireplaces. Equipment with an EER label will give the purchaser some objective standards for judging the cost efficiency of various units.

Installing a wood stove is not particularly difficult, but how it is installed is crucial. Whichever kind of stove is chosen, it should be installed according to the manufacturer's directions and in compliance with local building codes. In the past few years the number of deaths attributed directly to wood stove fires has risen alarmingly. In Maine alone, fifty-three persons were killed in a single year in stove-related fires, where in past years there were none. In Chesterfield, Virginia, a suburb of Richmond, 17 percent of residential fires in one heating season were attributed to wood stoves, and this figure no doubt holds for many other communities. These statistics will increase as millions of Americans use a heating technology they are no longer familiar with. It is essential that installers comply with recommended distances from combustible walls and floors, and use ventilated collars around stove pipes in walls and roofs. There are also problems with burning green or wet wood or maintaining too low combustion temperatures, both of which cause creosote buildups and increase the chances of destructive chimney fires. Anyone with a wood stove should most certainly install smoke detectors.

Aside from safety, anyone contemplating installing a wood stove should be aware that once the romance and nostalgia of this old-fashioned heating method have worn off, what is left is hauling wood, tending the fire, emptying ashes, and cleaning ash dust—chores that can quickly lose their fun. The cost efficiency of a wood stove is also closely related to an abundant supply of cheap, accessible firewood. All of these considerations must enter into an owner-builder's decision on whether or not to heat with wood.

SHEET ROCK SKILLS

If there is any job that I wished I had farmed out to a professional it is finishing dry wall, or as professionals usually call it, sheet rock. I am not alone in this; my survey of owner-builders confirmed my impression that finishing sheet rock—placing tape over the joints, troweling joint compound over it and sanding it smooth, three times—is one of the filthiest, trickiest, least satisfying jobs in all of house building. A large percentage considered it the most difficult job; virtually none thought it the easiest. Nailing up sheet rock is the essence of simplicity; taping and finishing it is a craft that requires practice, practice, practice. By the time I finished the last room in my house I had become a fair sheet rock finisher—never a good one—but you can read my progress on the walls of my house.

One reason why I and many other owner-builders did not have the sheet rock finished by a professional is because my house was never in a state of readiness to close in all of the walls. Unlike a contracted house, where each system is completed in sequence, the systems of an owner-built house are seldom all carried out to completion in the standard order. The flexibility of the owner-built method allows the builder to shift from one job to another at any time, leaving a system incomplete or delaying one while money, skill, or materials are accumulated. When an owner-builder moves into an incomplete house, as many of them do, building out of sequence is common, for it is then necessary to complete a bathroom, kitchen, and bedroom, at least to the point where they are functional. This throws the entire project out of sequence. In the case of sheet rock, I finished a part of one bathroom first, but did not finish the final room, the dining room, until nearly a year later. Working this way, it is obviously impossible to bring a professional in to finish a room or two at a time. The necessity for self-building those systems that are partially completed occurs frequently in a house that is lived in before it is finished. What might be called the nonsequential imperative—having to build yourself what you would rather subcontract—happens to just about all builders who do most of the work themselves.

Builders who do their own sheet rock finishing must be particularly careful on walls where a light fixture is mounted. The light will cast shadows on the slightest imperfection in the taping. I learned to ferret out these otherwise unseen irregularities by holding a work light against the wall when I sanded the final coat of joint compound. Another problem the owner-builder will encounter while doing sheet rock taping is the white dust which will cast a ghostly pallor over everything. I have heard more complaints about this part of the building sequence than any other. The dust is not only ubiquitous and unsightly; it also presents a danger to health. Every effort should be made to keep it out of the lungs by wearing a paint mask during sanding. One way of cutting down on sheet rock dust is to smooth the final coat with a damp sponge instead of sand paper. It goes more slowly, but does a clean job.

Of all of the house-building tasks described in do-it-yourself manuals, sheet rock finishing is the one I have found to be most inadequately treated, possibly because few of the writers have done it themselves. Virtually none of the manuals deals with the real problems involved—how to hold the trowel, the angle between trowel and wall, the amount of pressure used, and the consistency of the compound—all of which are crucial.

Finishing sheet rock, one of the more difficult skills for an owner-builder to master, is made easier by using a work light to cast shadows on imperfections.

One of the best descriptions I've found is in Dan Browne's *The House Building Book.* He has taught taping and finishing to students, and obviously knows the difficulties involved. Browne offers this pessimistic observation: "In San Francisco, my students and I took on a large renovation job, a part of which involved installing and taping a thousand pieces of sheet rock. After some instruction and three weeks of working steadily at compound and taping, not one person had achieved good quality or even a third the speed of a professional."

Taping and finishing sheet rock is another of those jobs that the owner-builder will learn more about by watching an expert at work than by any amount of reading. The difficulty in getting to observe professionals is that you have to catch them when they are doing the finish coat, usually the third. For the first two, they usually slap the compound on so quickly and carelessly that you will think there is nothing to it. However, for the final coat they use their considerable skill to ensure that the wall is perfectly smooth, and this is the time when one can learn something of value.

CARPENTRY SKILLS: WINDOWS, DOORS, AND CABINETS

Manufactured windows and prehung doors greatly simplify what used to be skilled carpentry jobs. When owner-builders install manufactured windows or doors, in effect they are purchasing the skills that were applied to the materials at the factory.

Windows are merely inserted into the rough openings, nailed into position, and finished with trim or molding. If economy is a consideration, the self-help builder should consider installing fixed windows wherever light, but not ventilation, is needed. These are constructed by fitting a piece of glass, usually ¼-inch plate unless the windows are quite small, into a box frame built into the rough opening. If double-glazed windows are desired, the rough opening can be constructed to take whatever standard double-glazed sizes are available. A favorite is a replacement panel for a sliding glass door.

Prehung doors must be carefully squared, but these too are

simply nailed into rough openings. It is possible to save money by constructing the door jamb, installing hinges, and hanging the door yourself. I did it both ways and found that the doors I hung myself fitted as well as or better than the prehung doors. Finished carpentry work requires a table or radial-arm saw, though, and takes time. Once a single door jamb is constructed and the door hung, the builder will be able to construct additional doors fairly quickly. Browne's *The House Building Book* gives a good description of how to hang doors from scratch.

Kitchen cabinets are one of the costliest items in the interior of a home because they represent a great deal of labor and skill. I found from my survey of owner-builders that very few build their own cabinets, but many install them. Purchased cabinets are designed to be installed easily; they fit together snugly, are available in a range of sizes, and adapt to any kitchen arrangement. Any owner-builder who has handled a hammer, drill, and screw driver will be able to install cabinets without difficulty.

Kitchen cabinets, counter tops, and ceramic tile work such as these in the author's home can be constructed by the average owner-builder—if time is available.

If time is available, I believe that kitchen cabinets are well worth learning to make. In constructing them by hand, the builder will not only save more money than on any other single item in the house, but will also learn a skill that can be put to use repeatedly for built-ins of every kind. Kitchen cabinets are essentially a series of boxes with drawers, shelves, and doors. Expensive, manufactured cabinets are fine pieces of furniture; they are fabricated with hardwoods, doweled and dovetailed joints, and baked finishes. Units made at cabinet shops are usually put together with a pneumatic nailer and simple glue joints. This is not to say they are not well made; on the contrary, they may function as well as much more expensive, manufactured units, depending on the workmanship of the particular shop. But an owner-builder, with practice, can make a cabinet as well as, and possibly better than, a cabinet shop.

The secret to making cabinets that are handsome and durable is to use good materials and to have a saw capable of making perfectly square cuts to a ⅟₃₂-inch tolerance. The material is readily available: ¾-inch birch veneer plywood. Birch takes any stain and produces a smooth, professional finish with a minimum of labor. Most cabinet shops use birch exclusively. If cabinets are to be painted, a less expensive grade of plywood can be used.

I built my own kitchen cabinets, and perhaps my experience will encourage fainthearted owner-builders to at least give it a try. I decided that I would attempt to build the smallest wall-hung cabinet I needed; if it didn't turn out to my satisfaction I could abandon the project, and all I would have lost was some plywood and my time. I had never before built anything as remotely ambitious as a cabinet, but then, I never before had had a radial-arm saw. I took my time, worked patiently, and completed a 1½- by 3-foot wall cabinet with two shelves and a door. Working part time, it took me nearly a week.

I was reasonably satisfied with the way it looked, however, so I built another just like it; and then I proceeded to construct all of the wall cabinets in the kitchen. By the time I began building the base cabinets I had gained enough confidence to convince myself that I could finish the kitchen. I was also working with ever-increasing speed and skill. Once I had learned how to make a decent-looking cabinet I ran amok and constructed built-ins in virtually every room in the house until my wife finally cried, "Enough!" Here are some of the things I learned about cabinet making.

When I began, I used dowels in all joints; but I discovered that, even using a good dowel jig and gluing and clamping all joints, I could not always get perfectly aligned joints. I finally abandoned dowels in favor of concealed screws drilled through blocks behind the cabinet stiles and in the corners. Screws and glue are every bit as strong as dowels and infinitely less time-consuming. One must remind oneself that it is the glue that holds the cabinet together, not the fasteners.

I purchased the laminated plastic countertops for my cabinets from a shop that specialized in custom-manufacturing them. The "continuous top" unit is molded under heat and pressure so that the backsplash and top are a single piece of plastic. This kind of unit cannot be made by the self-help builder, but he or she can construct units with separate top and backsplash. Installing countertops is tricky because the cement used is non-adjusting, and a router is needed to get perfect edges. I also installed a ceramic tile countertop on a kitchen bar and found this to be relatively easy. Tile countertops are now fashionable in many custom homes; they are an attractive and durable self-build alternative to plastic laminates.

The plywood edge of cabinets is most easily concealed by gluing onto the edge a strip of wood-tape that matches the plywood veneer. Take care not to get glue on the raw wood. The glue seals the wood so it will not take stain. (I found this to be a constant difficulty.) Strips of molding can also be nailed and glued to hide a plywood edge.

Cabinet doors and drawers can be simplified by allowing them to overlap the openings. Fitting doors and drawers into the openings with a rabbet cut—best made with a router—requires more accuracy. The kinds of hardware found on expensive cabinets are available to the amateur, including spring-loaded hinges and magnetic catches. Drawer slides are an important accessory. They are inexpensive, easy to install, and glide the drawer slickly into place.

One of the pleasures of building one's own cabinets is customizing drawers and shelves for appliances and cooking utensils. I richly enjoyed designing and building special drawers and storage spaces—putting to use my newly acquired cabinetry skills. The same skills were used in designing and constructing a built-in cabinet for my sound system.

PAINTING SKILLS

Watching a crew paint a house in a developer tract is a sobering experience for house builders contemplating painting their own homes. Inside, a worker, covered head to toe, face masked with a respirator, paint gun in hand, sprays a fog of white latex paint on walls, ceilings, windows, floors, and on himself. In the basement, another worker has draped a polyethylene sheet over a wall, has lined the house's entire complement of doors against it, and is spraying them with stain. Outside, wood trim stretched on sawhorses is being painted or perhaps siding is being stained—all with spray guns, their umbilicals attached to five-gallon cans. More than half of the paint or stain is lost in the wind, but time and labor are the expensive items in this operation.

This is the way mass-produced housing is painted. A crew of unskilled workers can cover a house, inside and out, with fast-drying latex paint in a day or two. Touch-up or trim painting, which must be done with a brush, is kept to an absolute minimum. An owner-builder can duplicate this technique with a borrowed or rented compressor and paint gun, but there is an alternative. It is not as efficient, but it gets the job done quickly and cleanly: the paint roller. A roller is a magnificent piece of intermediate technology; it is cheap, fast, and in most cases does a superior job to a brush. A child can learn to use it in no time and achieve professional results.

The advantage of the roller over the spray gun is that the self-help builder may not be ready to paint the entire house at once and may want to use a variety of colors. (The all-white interior that a house buyer moves into is often repainted with a roller almost immediately.) The roller offers complete flexibility, particularly to the builder who moves into a partially completed house and finishes it room by room.

Exterior siding is often purchased already painted or stained—

A paint roller is a valuable yet inexpensive aid, either for new work or for rehabbing an older house as this builder is doing. It enables a painter to do professional-looking work with a minimum of effort, and it is a significant time-saver as well.

If owner-builders do only one job themselves,
it will probably be painting, one of the easiest
and quickest of all the building skills. (Photo:
George Ballis)

an important saving in time and money—but if it is not, the amateur painter should seriously consider painting or staining it before affixing it to the house. This can be done with the siding laid flat on the ground or on saw horses, a simple, fast, and dripless method. I stained my 4 × 8 sheets of plywood siding before nailing them to the walls, and was thankful for it when I finished. I had only the trim to paint with a brush, and this kept my acrobatics on a ladder to a minimum. On the other hand, I failed to stain my open-beam, tongue-and-groove ceiling before applying the 2 × 6 boards, so this work had to be done with a brush. Since oil stain, unlike latex paint, is the consistency of water, it runs down your brush handle, hand, arm, body, and ultimately into your shoes as you apply it overhead. I bribed my son to do as much of it as he was willing to.

My survey of owner-builders did not surprise me when it disclosed that if self-help builders do only one job themselves, other than contracting the house, it will probably be painting. A large numbeer thought painting was the easiest of all the house systems, and quite a few included it with decorating as the most rewarding. One can only conclude that every self-help builder should paint his or her house; the skills are negligible, the results satisfying, the savings enormous.

FLOOR COVERING SKILLS

The Census Bureau decrees that a house is officially completed when its floor covering is installed. Leaving the floor covering until last makes obvious sense, because just about every other system deposits litter, dust, dirt, scrapes, scratches, dents, as well as spills of every conceivable kind on the floor. By the time a builder has mastered the ceiling and wall systems, there should be no reluctance to take on the floors. There is a great variety of materials available for floor covering, and all but a few of them present no more difficulties than have been encountered in other systems. Hardwood floors are perhaps the toughest to install. Oak flooring, the most common, should be installed with a floor nailing machine; without one, this task can be a frustrating finger banger.

Resilient floor coverings such as vinyl are perfectly suited to the amateur. They are laid in either tiles or sheets over a coating of adhesive cement applied directly to the subfloor. Aside from a certain amount of care in handling and cutting this thin material, installing it takes only time and patience. The floor covering industry, seeing a market in owner-installed materials, has produced a series of vinyl and carpet tiles with adhesive backings; the backing is peeled off and the tiles are simply pressed in place on the floor. Since they do not compare, by any stretch of the imagination, with one-piece flooring installations, a builder would be wise to take a look at a floor laid this way before deciding on it.

Ceramic tiles are often laid in bathroom floors. These can now be installed on a base of adhesive cement, rather than on a masonry base as was once necessary. With a bit of planning, and the use of a tile cutter, ceramic tiles are quite manageable.

Wall-to-wall carpeting remains the most popular floor covering in America. Aside from kitchens and baths, most houses are carpeted throughout. (This may change, however; passive solar houses require concrete or masonry floors for thermal mass.) Installing wall-to-wall carpeting so that it is properly seamed and stretched is a ticklish job. It can be done by the amateur, and I've seen owner-builder carpeting that was quite professional looking. But I've also seen some that waved and wrinkled and displayed its seams like earth faults.

I chose to have my carpet professionally installed, for several reasons. Carpeting is regularly discounted heavily by department, home furnishing, and floor covering stores. One can often get carpet installed for the same price another retailer sells it uninstalled. With judicious shopping, installation can be relatively cheap. Installing carpet properly takes a kick bar, seamer, and a considerable amount of skill. It is something like sheet rock in this respect, but unlike dry wall finishing, carpeting is installed rapidly and cleanly by those who do it for a living. In the morning your house is unfinished; in the evening it is magically carpeted—complete. That experience of having the final system finished for you in a day is a rather nice way of rewarding yourself for all of those jobs that took so, so much longer.

FOUR

STICKS AND STONES:
THE STUFF OF BUILDING

n theory, people who build their own homes have complete freedom to choose the materials that are used. In practice, this is seldom the case. Choices are often subtly dictated by the section of the country one lives in; by the community one is building in; by the tastes of fellow workers, friends, or family; by energy considerations unthought of a few years ago; and by personal needs one is often only dimly aware of. Ignorance about the project that is being launched and insecurity about a lack of expertise also strongly affect the decisions made about building materials. A builder is more likely to choose to construct a house with those materials he or she feels will be easiest to work with. In most cases this will be wood, the material already used if any kind of amateur construction work has been done. Almost everyone has sawed a board and hammered a nail; unlike laying a brick, say, there is little mystery to it.

WOOD

Aside from adobe or concrete block houses, all standard house construction is platform framed or "stick built." A skeleton is erected of dimensional lumber, mostly 2 × 4's, 2 × 6's, and 2 × 8's, and over this frame are layered the floor, roof decking, and wall materials. Almost every owner-builder constructs a good portion of the house of wood, regardless of what is chosen for the roofing and outside walls.

Wood is one of the most satisfying of all materials to build with. It is delivered to the building site fresh from the mill, the smell of sawdust and resin still on it, cut into the exact geometries of the building science. It is called lumber, but the term denies wood its roots: these are trees that have been sliced with Euclidian precision into floor joists, studs, and rafters. The heritage of the forest is not lost, for knots and grain are reminders of recent limbs and annual growth. Handling these lengths of julienned tree, smoothed and squared by the sawmiller's art, brings builders into contact with the same elemental forces from which sprang their need for shelter. Their houses are canopied forests, transmogrified into conventional shapes that have evolved from centuries of building by millions of builders.

Plywood transforms the tree even more radically; it is shaved into wooden blankets, sandwiched, glued, and pressed, then issued in slender, cross-grained slabs rendered incredibly strong and rigid by the manufacturer's skill. Because it is so severely altered from its living form, plywood retains less of the mark of the tree, but when cut it bleeds sawdust. Like all dried and seasoned woods, its fibers clasp a nail as tenaciously as jaws.

Hardwood veneer plywood is an absolute joy to work with. Whereas construction grade plywood is rough and splintery, with almond-shaped knothole patches, birch or (heaven forbid the cost) walnut, cherry, or oak veneer plywoods are smooth and fine-grained. One handles them like plate glass and cuts them with reverence.

Wood impregnated with chemicals to preserve it from termites or rot is much less attractive to use. I have cut into pressure-treated 2 × 4's that spattered juice like a sawed orange. Yet, treated lumber has found its way increasingly into house construction. It is invariably used for decks, and special grades of treated lumber are being used underground for basements and foundations. This may sound preposterous—burying under the earth a material that traditionally must be bone dry to resist rotting away like compost—but the copper arsenate preservatives pumped into the fibers actually eliminate dry rot and repel insects. It has been installed in thousands of houses and is approved by HUD and FHA. Life estimates exceed 100 years. It allows owner-builders to construct a basement or crawl space foundation using the skill they know best—carpentry. Sections of foundation can be constructed like wall sections, flat on the ground, and tilted into position on a bed of gravel. An all-wood foundation system makes it possible for the builder to finish a house without laying a single brick or concrete block or pouring a foot of concrete.

Other advantages of this system are that walls below grade can be insulated, electrical wiring can be snaked through, and such standard interior wall finishes as sheet rock or paneling can be used in basements. This system can also be installed in the winter when concrete cannot be poured because of the cold. Lumber to be used underground is not the same as treated lumber used for decks and outdoor construction, however; it is a special grade and sells at a premium price. Any decision on whether to build with a wood foundation will have to depend on such individual considerations as the availability of treated lumber at a price competitive with masonry and how comfortable the self-help builder feels about carpentry as opposed to masonry or concrete. Solar design could also play an important part in such a decision. Where the crawl space or basement acts as a heat storage element in a convective loop it may be essential to insulate the foundation walls. A wood system with six-inch studs offers maximum below-grade insulation. Those wishing to explore wood foundations should consult the OVE manual.

The decision on whether or not an owner-builder uses wood siding will probably depend as much on the part of the country he or she lives in as on any personal considerations. The Census Bureau keeps statistics on the kind of exterior wall materials owner-builders place on their homes. In the Northeast and North central parts of the country, 65 to 70 percent of all owner-built homes have wood or wood products siding. (The most common wood product siding is hardboard, a manufactured wood fiber material that is economical, durable, and is easily sawed, nailed, and painted. It is sold under such brand names as Masonite and Abitibi, and is available in a wide variety of sizes and finishes.) In the West, 60 percent use wood, and in the South, only 35 percent. These statistics reflect long-standing building traditions derived from historical considerations.

When the colonists arrived in this country in the seventeenth century they found climatic and environmental conditions quite different from those they had experienced in England: temperature extremes were far greater, with colder winters and warmer summers. Their resources for housing themselves in this new environment were an abundance of forests and a wood-frame technology they brought with them from England. The colonists quickly adapted a housing system suited to the needs of the harsh New England climate. With an ingenuity that was characteristic of the Puritans, they mechanized the cutting of lumber. The first sawmill, at Pescataqua Falls, between Maine and New Hampshire, was erected only thirteen years after the Pilgrims landed. Long before sawmills appeared in England, the colonists developed a technology for producing board lumber quickly and cheaply. The result was the wood-framed, clapboard structure so familiar in the Northeast—the saltbox house.

By the start of the nineteenth century, the lumber industry, now powered by steam, was turning out vast quantities of wood products for inexpensive, durable, domestic houses. By the mid-nineteenth century, with the development of balloon framing and the inexpensive, manufactured nail, wood was firmly established as the preeminent house construction material throughout the nation, but particularly in the North and North central states. (For a review of the early development of American housing see James Marston Fitch, *American Building 1: The Historical Forces That Shaped It*, 2nd ed., rev., Boston: Houghton Mifflin, 1966.)

That wood should remain the material of choice is yet another example of cultural inertia, the persistence with which social behavior runs in well worn grooves. Although I have argued that individual builders re-create themselves in their shelter, the individual is also influenced by social pressures to conform to local construction practices. Your image of what a house should look like is determined, not only by private psychological needs, but by the homes of your neighbors and friends. The builder who chooses materials that are different from those considered "standard" by neighbors always risks criticism. The more socially cohesive the community, the more strict the demands for conformity. In a modern metropolitan suburban community one will find in a single block, houses with exteriors of wood, brick, stucco, aluminum, and vinyl, but the owners of those houses may scarcely know each other. Self-help builders constructing homes in small towns, however, may find much more uniformity of style in surrounding houses, in keeping with the tight social fabric of the community. As a result, choices may, consciously or unconsciously, be constrained by how closely the builder identifies with the social values of neighbors.

There are a large number of owner-builders, however, who consciously reject the standard image of a modern American home. They build in a style that is a throwback to a romantic past, to a tradition rich in myth and nostalgia—the log cabin. A log home would seem to be a perverse anachronism; it regresses to a housing technology that is inefficient in the use of materials, hard to build, and difficult to waterproof and insulate. Yet it retains its popularity—a fact supported by the dozens of manufacturers of packaged log homes that have appeared in recent years.

Constructing a modern house of logs that you have cut and processed yourself is a herculean task that will tax the strength of all but the most dedicated owner-builders. Trees must be felled, branches pruned, logs hauled to the site, bark stripped, and logs dried and treated with wood preservative. The logs have to be cut to size, notched, and mortised or drilled for spikes. Then they are lifted into position, usually with a block and tackle. All of this merely gets you four walls. It is little wonder that few builders are willing to attempt building a log home from available timber on their land, even if the lure of free building materials is seductive.

The manufactured log house has made it possible for many Americans to consummate their love affairs with logs. A packaged log home comes to the builder with the entire log wall system prepared for assembly. Logs are stripped, leveled on two sides, notched, treated, and cut to exact size. Most log houses are sold as shells; builders must finish the interior sys-

This hewn-timber house from the nineteenth century is a fine example of the timber-wall technology that is responsible for the "log cabin myth."

Modern log home kits, such as the one under construction, have made possible the consummation of America's love affair with logs.

The charm and attraction of the log home is shown in the warm, rustic atmosphere of the interior of an owner-built package log home.

tems with materials they have purchased themselves. Therefore, a package log house usually does not represent any savings
in time or effort over a package house of standard platform
frame construction. In fact, if the builder puts a stud wall behind
the exterior log wall to get extra insulation, the work and cost
could be considerably more than standard stud walls with wood
siding.

What then is the appeal of the log house? The reply to this
question must be speculative, but there are cultural, historical,
and psychological clues. For most Americans, two kinds of
houses typify our national heritage, the Williamsburg colonial
and the log cabin. For generations, schoolchildren have been
edified by the log cabin myth. Perhaps an 1840 campaign oration
by Daniel Webster—describing his family's New Hampshire
log home (which he had the misfortune *not* to have been born
in)—expresses this myth most sentimentally:

> I carry my children to it, to teach them the hardships endured
> by generations which have gone before them. I love to dwell
> on the tender recollections, the kindred ties, the early affec-
> tions, and the touching narratives and incidents, which mingle
> with all I know of this primitive family abode. I weep to think
> that none of those who inhabited it are now among the living;
> and if ever I am ashamed of it, or if I ever fail in affectionate
> veneration for him who reared it, and defended it against savage
> violence and destruction, cherished all the domestic virtues
> beneath its roof, and through the fire and blood of a seven
> years' revolutionary war, shrunk from no danger, no toil, no
> sacrifice, to serve his country, and to raise his children to a
> condition better than his own, may my name and the name
> of my posterity be blotted for ever from the memory of man-
> kind! (*Works, II*, 1851)

Webster's grandiloquence was inspired by a log home in the
Northeast, but for most Americans the log cabin is associated
with the frontier as a symbol of independence and determination in winning the West. The frontiersman and his family
have been depicted in thousands of films and TV shows; they
are invariably protected from the hostilities of weather and
Indian by a thick wall of timber. We are inspired by Abe

Lincoln, who rose from a humble log cabin to the White House. We eat pancakes and Log Cabin syrup, and build toy forts and stockades with Lincoln Logs.

All of this has undoubtedly contributed to a nostalgia for the log home that runs very deep in the American spirit. In many cases, builders of log homes re-create a fantasy life that trails back into the dim reaches of childhood, to pleasures that have long since dropped from active memory. Other log builders may be searching for a romantic escape from the machine-made conformity of modern housing. You could argue that this escape is an illusion, that the log house is as fake as the colonial, Tudor, or Spanish style houses that may stand on the same street with it, because the style is, to use Frank Lloyd Wright's expression, not organic to the structure. It is a contemporary house with a primitive log exterior overlaid upon it, just as a Williamsburg style house is the same basic housing system, perhaps the same floor plan, with a colonial veneer attached to it.

There is a certain truth to this, but the log home does use a material—raw logs—that defines its structure inside and out. The word most log builders use to describe this material is "rustic." The expression suggests a rejection of the urban environment and everything it stands for in favor of a return to the woods as a source of inspiration for one's dwelling, even if the finished house has all the modern conveniences of any other suburban dwelling. The knots, checks, and splits are a palpable reminder of the living tree which once offered temporary shelter, but which now provides permanent protection from the elements. As one log builder commented: "As I sit in front of my fireplace and look around at the log walls and up at the log beams I feel like I'm enclosed by a tight forest that protects me from whatever is out there to do me harm."

There are certain caveats about packaged log houses that should be mentioned. The cost of a finished log home comes to approximately three times the package price. This fact is often concealed by log home salesmen whose aggressive selling practices help account for the popularity of the packaged house. The distance from the factory is important to costs, because shipping can account for a sizable portion of the total package price. Building codes may present difficulties because codes are based on standard frame construction, and a log house may not meet codes. Anyone living in an area where codes are enforced should check them before seriously considering a log dwelling. It is also absolutely essential with a log house that the foundation be square; logs are not as forgiving of miscalculations as is a frame structure.

The usual precautions taken with any manufactured home must also be taken with log packages. Be certain that all of the logs are delivered with the package; you cannot run down to the local lumberyard and buy a couple of replacements. Make sure a few extra logs are included in the package. It is inevitable that some logs will be warped, badly checked, or unusable; sometimes logs are not properly marked or are improperly cut. Replacing them will cause lengthy delays because they must be shipped from the factory. Don't work with logs that are not properly dried—warping and shrinking could prove to be a serious problem. Log houses are inherently difficult to waterproof and insulate, and inferior materials magnify these problems. Finally, since wood-boring insects are more of a problem with log homes than with other kinds of construction, the logs must be treated periodically with a chemical preservative.

Another kind of packaged wooden house that utilizes a distinctive building material is the all-cedar home. There are a number of manufacturers of precut cedar houses, most of them in the Pacific Northwest where cedar forests are abundant. Cedar houses are usually open-beam, a construction method which displays the timbers and decking of the attractive cedar wood. The Swiss chalet is a popular style with cedar manufacturers.

Cedar is an excellent wood for house construction because, like redwood, it is naturally resistant to insect intrusion. If left unstained, it weathers to a handsome silver-gray, characteristic of many cedar-shake roofs. Cedar is also an attractive and durable wood for interior cabinets.

Cedar home manufacturers have been hard hit by the energy crunch. There are two reasons: increased transportation costs have made the shipment of materials from the Northwest to other parts of the country extremely expensive; and many of the cedar home systems were developed without serious consideration for insulation. The selling point of all-cedar construction is putting the wood where it can be seen; instead of using sheet rock on interior walls and ceilings, tongue-and-groove cedar boards are used for both exterior and interior walls and roof decking. With proper insulation, this is an expensive construction system; some manufacturers have attempted to cut costs by developing laminated wall systems which combine strength with insulating efficiency.

Unless an owner-builder is infatuated with cedar, a packaged cedar home makes little economic sense in many parts of the country. There is little that you can do with cedar construction that you cannot do with pine or fir; and with the proper stains, the builder can closely approximate the look of cedar. Blueprints for the chalet and barn-style houses that cedar packagers favor are readily available from house plan services and allow for a great deal more flexibility of design and construction than does a package.

BRICK

Regional differences in the preferences owner-builders show for building materials are perhaps no more sharply delineated than in the use of brick for the exterior of a house. Census Bureau figures show that in the Northeast, Central, and Western parts of the nation approximately 10 percent of all self-built houses are faced with brick. In the South, 55 percent are brick. How is this wide discrepancy explained?

I don't believe there is a simple explanation for the overwhelming preference in the South for brick over any other kind of building material. In the previous chapter I noted that only a small percentage of self-help builders are willing to do brick work themselves; the large majority who face their houses with brick veneer subcontract the work to a professional mason. This work can amount to a sizable piece of the overall cost of the house. Labor costs have traditionally been much lower in the nonunion South, so it is possible to have professional masonry for a good deal less money.

Thomas Jefferson's Monticello was a typical owner-built home—it took forty years to complete! Jefferson's mansion, and other plantation houses like it, established brick as the most desired construction material in the South. (Photo: Library of Congress)

Economics are, therefore, a consideration, but not a major one. This same preference for brick is shown in contractor and developer-built houses in the South, although to a lesser degree than in self-built homes. As with the choice of wood exteriors in the Northeast, historical reasons offer a much more persuasive explanation. In the South, brick is associated with quality construction, wood with substandard building. When my mother-in-law in Louisiana learned that my owner-built house had wood siding, her estimation of it fell like the bricks it should have been built with. Low-income housing in the rural South has always been wood frame; public buildings and the homes of the affluent were of brick, like the plantation houses of the Southern aristocracy.

I paid a recent visit to Thomas Jefferson's magnificent owner-built home. Monticello was constructed over a period of forty years. Its walls are brick—sixteen inches thick—and the brick was manufactured on Jefferson's estate. Monticello served as a model for the homes of generations of Southern planters. Behind the stately columned porticoes and white plastered exteriors of the ante-bellum mansions of the South are thick walls of brick. The tradition still lives: Southerners consider brick "quality," and Southern self-help builders are willing to pay whatever it costs to dress their homes in masonry finery.

One historical reason for the prejudice toward brick in the North is worth mentioning. Climate always plays a role in the selection of building materials, and there were climatic reasons for the rejection of brick in colonial New England. Before vapor barriers were used, northern regions had problems with condensation on the interior brickwork. The colonists believed this "weeping" caused disease and death from "malignant vapours." Jefferson explained the phenomenon of condensation and prevented it at Monticello by waterproofing the exterior brick walls with pine oil. In the South, condensation was not a serious problem because of the lack of temperature extremes. Climate plays no part in condensation control in modern brick cavity wall construction, of course. But building patterns, once established, possess a powerful cultural inertia.

Much of the appeal of a brick house is that it is virtually maintenance free. Wood exteriors must be painted every few years, whereas brick needs no such attention. It ages to a fine patina which looks better with time. A brick house is much more competitive with wood exteriors when maintenance costs over a number of years are calculated in the cost equation. The problem with this kind of long-term cost evaluation, however, is that most self-help builders construct their houses within a tightly limited budget and they are concerned only with immediate costs. My sample of owner-builders shows that many of them intend to build another house in the future; they are not necessarily building for posterity, and long-term maintenance is not a major concern.

The minority of owner-builders who do brick masonry themselves and who enjoy it probably share certain personality characteristics. Planning and laying up a brick wall is an exercise in geometric certainties. There is no margin for error in brick work; each row must come out even, each brick must fit its niche in the wall perfectly. Nor is it a task for the sloppy, devil-may-care workman; cleanliness is essential for a decent-looking brick wall. Taking pleasure in small-scale, close work; feeling comfortable with repetitive tasks; patience and accuracy; and an appreciation for the detail within the larger pattern—these are the personality traits of the ideal amateur brick mason. He or she is not unlike the person who enjoys making models, doing hand work such as needlepoint, or modifying a car engine. Without at least some of these characteristics, those usually identified with the obsessive-compulsive personality, an amateur mason could find bricklaying to be a burdensome, frustrating millstone.

STONE

Very few owner-builders construct stone houses for the same reason that few build log houses from raw lumber—the work in both cases is oppressive. The hero and heroine of stone construction are Scott and Helen Nearing, who built a series of stone buildings in Vermont and Maine and whose books are biblical for many counterculture builders. The Nearings improved upon an earlier system for setting up rock walls in movable forms. The Nearings fill the forms with rock and poured concrete and are able to build attractive stone buildings in a fraction of the time it would take to lay up a stone wall rock by rock. Anyone who has a yen to construct a house of stone— a material which is usually free for the hauling—should consult the Nearings' books for information and inspiration. Their final word on stone construction is: "People of moderate intelligence, little experience, and slender means can build with stone if they have the time, patience, and the inclination."

Most owner-builders who work with stone use it for part of the exterior veneer or for a fireplace. Whereas the regularities of brick present problems of mathematics and geometry, the irregularities of stone must be resolved by insight and intuition. For the ingenious terpitect, placing rocks into a perfectly integrated pattern of size and fit can be highly creative and uncommonly satisfying.

I built a fourteen-foot stone veneer fireplace, and my experiences are probably fairly typical of the pleasures and difficulties of working with this recalcitrant material. I had culled a pile of fieldstone from the grading of my building site, and I quickly exhausted these in constructing the hearth and a few feet of my fireplace. I found a supply of stone at a road construction site about ten miles away, so I made about a dozen trips with my Volkswagen van, hauling rocks that I had selected for their flatness and suitable size. This sounds matter-of-fact, but I recall searching a roadbed for just the right stones, carrying them individually—hundreds of them—more than a block under the summer sun to the van—exhausting work.

At my building site, the rocks were scrubbed with water and a wire brush, placed in a wheelbarrow and carried up ramps to the living room where they were spread out over the floor like so many recruits awaiting induction into the permanent rites of mortardom. Each candidate for a position in the fireplace was lifted to a six-foot scaffold and tentatively fitted into its allotted niche. If it fit closely but not perfectly, I chipped at its edges with a mason's hammer until it eased snugly into its permanent position. I then removed it, slathered a thick bed of mortar against the brick backing, and pressed the stone into final place. I could ordinarily do no more than a single horizontal row each day, two at the most, because until the mortar hardened, the stones were flattened precariously against the bricks and placing additional weight on top of them would cause the bottom rocks to avalanche to the floor—an unsettling event that occurred on several occasions. This meant that the work progressed very slowly; in fact, it took me several months of part-time work to finish the fireplace.

In spite of this slow and laborious process, I found that building my stone fireplace was satisfying. Perhaps it was because it involved the hearth, the central core of a house. Perhaps it was the aesthetics of creating a pleasing pattern of stone from the motley array of rocks spread out on my living room floor. I'm inclined to think it was both.

ADOBE

Not long ago, my wife and I were traveling through New Mexico when we drove through a small town tucked into a valley in the Sangre de Cristo Mountains. I stopped by a construction site alongside the road, an adobe house with no more than about 800 square feet of living space. A stack of adobe blocks were curing in a corner of the house—New Mexico mud mixed with straw and formed into blocks 12 × 18 × 4 inches. In front of the house were several dozen stripped pine logs obtained from the nearby Kit Carson National Forest, free for the cutting and hauling. These were the *vigas,* or roof beams, used in adobe construction for thousands of years. The floor was a parquet of rough-cut 2 × 6's that obviously were obtained as scrap from a lumber mill. As I looked at that construction site, the only thing I could see that the owner had paid for were concrete and a single row of concrete blocks for the footings. To this he was going to have to add windows, doors, plumbing, and electrical wiring, but otherwise he was building a dwelling for virtually nothing but his labor.

Not too far from this small, unfinished house I visited the site of another owner-built adobe. It was a 3,000-square-foot, four-bedroom house constructed around a center-court atrium. It was complete with several fireplaces, modern kitchen, dark-

One of the oldest owner-built communities in the nation is Taos Pueblo, an adobe settlement that has stood at the foot of the Sangre de Christo mountains in New Mexico for more than 700 years. Adobe construction is indigenous to much of the Southwest. (Photo: National Archives)

In the Southwest adobe is an ideal construction material for the owner-builder. Working with heavy adobe blocks and plastering walls is hard work, but the material is perfectly adapted to the hot, dry climate. (Photo: Library of Congress)

room—a modern house in every way. Yet it shared with the small, primitive adobe exactly the same construction: mud-straw adobe blocks plastered inside and out with the same adobe mud from which the bricks were made; it had pine vigas for rafters and an adobe roof with modern waterproofing. In one way it was even more traditional than the smaller house: the earth floors were to be soaked with ox blood to form a thick, tough surface that could be buffed to a handsome burgundy luster. At Taos Pueblo you can find the same kind of floors in adobes that are a thousand years old.

In the Southwest, adobe is an ideal construction material for the owner-builder. It is perfectly adapted for a dry climate with hot days and cool nights. The thick walls collect and store heat in the winter, and in the summer they are cooled at night to create a comfortable interior during most of the day. Adobe is also ideally suited for heavy, thermal-mass passive solar construction. Thrifty builders can make their own blocks, or they can purchase relatively inexpensive blocks in the areas where they are manufactured. If they must be shipped, however, their weight makes them a costly building material. That same weight also makes adobe blocks a material for the strong-backed; laying them at any height above waist level is wearing work. Those interested in exploring adobe and such forms of earthern construction as pressed brick and rammed earth should consult Kern's *The Owner-Built Home*. Kern, himself a mason, offers some of the soundest advice to be found on masonry buildings of every kind.

METALS: STEEL AND ALUMINUM

Steel, the most common building material in nondomestic construction, seldom finds its way into an owner-built house. Aside from steel I-beams used as foundation girders in difficult building sites or reinforcing steel used in concrete slabs, footings or masonry work, the strength of steel is seldom required in domestic construction. Steel does not compete in price with wood, and for most applications wood is prefectly adequate for structural strength. There are times, however, when steel can solve difficult problems of support or structural weakness, or simply add a measure of safety, but most self-help builders avoid steel because it is an unfamiliar material and because they have neither the necessary welding and cutting tools nor skills to work with steel. Steel supply houses stock a wide variety of commercial construction materials easily adapted to home building. They will cut and drill steel to a builder's specifications. It is a good idea for an owner-builder to shop around such an establishment to see what kinds of steel products are available, since many of them do not find their way into the building supply stores he or she normally does business with.

Galvanized steel sheet metal is particularly well suited for high-risk fire areas, and is still used in farm buildings and outbuildings as a roofing material. It is not often found in house construction. (The Nearings used it on their stone house in Vermont, however, and swear by it.) Owner-builders are most likely to use sheet galvanized steel for gutters, chimney flashing, and for forced-air heating and cooling systems. I did not particularly enjoy working with sheet metal, and I don't think my response to this material is unusual. I've never found a self-help builder who admitted deriving any great pleasure from sheet metal work of any kind, steel or aluminum. It is a material that lacks strength, malleability, plasticity, or solidity. It is an unforgiving material; once pinched or creased it cannot be returned easily to its original shape. Its edges are sharp and cutting, and because of lack of rigidity, it is terribly difficult to fasten sheet metal to anything, especially to another piece of sheet metal.

Aluminum has increasingly found its way into home construction in the form of siding. Aluminum siding offers one important advantage: it is virtually maintenance free. Paint finishes are baked on at the factory, and if properly installed, aluminum siding is almost indistinguishable in appearance from painted, horizontal wood-lap siding. Census Bureau figures record the increasing use of aluminum siding in all house construction; it now appears on 12 percent of all single-family houses built in the country.

Like other exterior materials in houses, aluminum siding shows geographical bias. It is highly popular in the Northeast and Central parts of the nation where it is used on at least one-fourth of all houses built, but is much less popular in the South and West where it appears on fewer than 5 percent of all houses. The reason for this sharp contrast is connected in part with the historical preference for horizontal wood-lap siding in the North and Midwest. Aluminum siding imitates wood, and is easily installed over existing wood siding in older houses. In the West and South, where brick and stucco are the preferred exterior house materials, aluminum is much less popular.

Fewer owner-built homes use aluminum siding than contractor or developer-built houses. One reason for this is that aluminum siding is aggressively marketed to homeowners by specialized contractors who offer installation as part of a fixed-purchase package. It has been much less energetically merchandised at building supply centers where the owner-builder makes his or her decisions on materials. There is little skill involved in the installation of aluminum siding; the system, once leveled and started, goes up much like a shingled roof. The material is nailed to the house frame and is easily cut with metal cutters or a circular saw. There is little reason for owner-builders to hesitate to install aluminum siding themselves, other than the cultural inertia which accounts for a time lag in the acceptance of any new material. Two objections to aluminum siding—it is noisy and gives off sheet metal sounds, and it offers no insulation value—have been countered by the development of siding with either fiberboard or polystyrene insulation materials bonded to the aluminum panels.

Many owner-builders turn to aluminum doors and windows because of their low cost and low maintenance, although they are not as aesthetically pleasing as wood. Aluminum is also less energy-efficient. Aluminum drain gutters offer low maintenance. However, aluminum does not hold paint well. Baked aluminum gutters and pipes are a solution to the painting difficulty but they are available only in white. The other major use of aluminum in house construction is in electrical wiring. Codes now place rigorous restrictions on aluminum wiring because of the danger of fire caused by poor connections. Aluminum can oxidize at outlet and service box connections, and the resulting resistance can cause overheating. Because of its cost advantage, aluminum is still widely used for heavy cables, but copper is universally used for general house wiring.

PLASTICS

We live in a world that is increasingly plastic; almost every manufactured product incorporates plastic parts, and thousands of the items that are everyday fixtures in our lives are entirely plastic. It is little wonder that many house systems previously fabricated from such traditional materials as masonry, wood, and metals are now made of plastic. There are many critics of the plastic revolution; most reject it on both ecological and aesthetic grounds. Plastics are manufactured from the by-products of petroleum and are therefore a scarce, nonrenewable resource. Many counterculture self-help builders argue forcefully that building your own shelter presents you with choices the home buyer does not have, and these choices should be on the side of the environment: wood, earth, stone, concrete.

These are low-technology materials that use energy efficiently, particularly if they come from near the building site. This argument is a call for social responsibility on the part of autonomous builders who have it within their power to choose to live in harmony with the earth. You could question whether selecting a wooden or ceramic countertop instead of a laminated plastic top, say, really matters all that much, but if you are committed to environmentalism, living with the nonplastic top will be a constant reminder of your choice and therefore a renewable source of pleasure and satisfaction.

The aesthetic argument is that plastics are for the most part cheap imitations of natural materials and that they are tasteless and ugly. No doubt, much of what is rendered into plastic is kitsch; these products often are not durable and not particularly well designed. As a building material, however, plastic has the potential for creative design and use that any other material has. I have, for example, two one-piece tub/shower units of fiberglass-acrylic in my house, and I find them every bit as aesthetically pleasing as the traditional porcelain tub sur-

Molded fiberglass and acrylic showers and bathtubs are replacing steel and tile tubs and showers in many owner-constructed homes. They are easily installed and require no special skills or equipment.

rounded by ceramic tile. They are well-designed, durable, easy to clean, and much more simply installed than a traditional unit. I also used some plastic molding that is an imitation of stained and painted wooden molding. To an ecologist, this is an unforgivable substitution, but I confess that I am no purist in such matters; and the moldings, which cannot be distinguished from wood except by close examination, saved me a great deal of time when time was important to me.

The aesthetic difference between the tub/shower and the plastic molding is that one uses plasticity, the ability to be molded, in a creative way that is unique to that material, whereas the other denies its individuality in an abject imitation of another material. Plastic laminated cabinets, furniture, and tabletops, for example, are manufactured both ways, in finishes that are unique to the material and in wood-grain and stone-finish imitations. In my view, the material is aesthetically more pleasing when it is honest, when it exploits its own inherent possibilities.

BRACE, BEVEL, AND BOB: THE HOUSEBUILDER'S TOOLS

magine yourself as a modern Crusoe, cast upon a desert island by a shipwreck. Washed up on the shore was the wreckage of the ship, wooden crates of every kind—a vast supply of usable wood. And lo, in one of the crates there was even a saw and a supply of nails—everything to make yourself a solid, durable shelter. But there was no hammer. How would you proceed to build without this single, commonplace tool? Yes, you could use a rock. But have you ever tried driving a nail with a rock? You would be compelled, before you attempted to construct anything from that supply of lumber, to fashion yourself some kind of primitive hammer—a rock lashed to a split wooden handle, perhaps. But how would you pull out a bent nail? This situation would be even worse if you had a supply of screws and no screwdriver.

One cannot build without tools. But many owner-builders, in working out a budget for their construction enterprise, fail to account for the very real expense of purchasing tools. Of course, most house builders begin their construction projects with a supply of tools on hand. If an owner-builder has constructed anything, he or she has probably accumulated an assortment of hand and power tools. In building a house, however, specialized tools become either necessary or particularly useful and time-saving.

Perhaps one of the best pieces of advice about tools I could offer to a potential house builder is purchase them for a particular work sequence before the work is started. This may seem so self-evident that only an idiot would fail to observe it; but I, and many self-help builders I have talked to, have made the mistake of starting a building system with makeshift or inadequate tools, only to purchase the proper tools when the work was nearly completed. The tools this is most likely to occur with are the expensive ones such as a table, chain, or radial-arm saw, or a concrete mixer.

My own experience with a concrete mixer is typical. I constructed a full concrete-block basement for my house. When I started, I mixed concrete and mortar with a hoe in a wheelbarrow. This is a slow, laborious way of mixing, but I was working alone and thought that it would suffice for this one masonry job. By the time I was two-thirds finished with my basement, the wheelbarrow system was becoming increasingly cumbersome, so I purchased a small, electric-powered concrete mixer and discovered immediately how much labor was saved and how much superior a mixing job it did. The advantages became more obvious when my son joined me in laying blocks and we needed mortar more quickly. I deeply regretted not buying the mixer at the start of the work and vowed not to make this same mistake again. My one consolation was that I at least had the wit to purchase a contractor's wheelbarrow— something every owner-builder ought to have—at the start of my masonry work.

With the lesson from the concrete mixer fresh in my mind, I purchased a radial-arm saw before I started framing my house. This tool enabled me to do more professional-looking carpentry than I ever dreamed possible. I am not alone in my admiration for this tool; in my survey of owner-builders I asked which tool was the most helpful in building a house. Next to a portable circular saw—perhaps the single essential power tool—a table saw or radial-arm saw was most frequently named. (Of the two, the radial-arm saw is the more versatile.) These tools give to an amateur the same capabilities for accuracy, dependability, and fine workmanship that a professional carpenter has. The only reason I was encouraged to build kitchen cabinets was because of the way my radial-arm saw produced such smooth-edged, splinterless rectangles from ¾-inch plywood.

The portable circular saw is the workhorse of home building, particularly framing; quite a number of self-help builders construct their homes with only this one power tool and a set of hand tools. Consider buying two circular saws for convenience: one lightweight, six-inch saw for cutting plywood sheets and light dimensional lumber, and one eight- or nine-inch heavy-duty saw for cross cutting heavy dimensional lumber. A table saw, however, can take the place of the heavy-duty saw. Having extra sets of a number of tools is often necessary if a builder is to have friends help out with carpentry work. An extra framing hammer, carpenter's apron, and tape measure are the bare minimum for each helper who is doing rough carpentry. Helpers can sometimes be counted on to bring their own tools, but

With a radial-arm saw I was able to perform feats of carpentry undreamed of in all my philosophies. Most owner-builders find that a table saw or radial-arm saw makes it possible to approximate professional cabinet work.

not always. Tools also have a way of being misplaced just when you need them, and an extra hammer, screwdriver, nail puller, tape measure, or carpenter's square often saves a great deal of frustration. When you are on the roof and the tool you need is in someone else's hand, you often wish for more than one. A good carpenter's apron can eliminate much of the problem of lost tools if you make it a point of honor always to return each tool to its pocket on the apron. Unfortunately, such discipline must come with many years of practice; I never acquired it.

Another power tool frequently mentioned by owner-builders as especially useful is the sabre saw. Like a power drill, a sabre saw can be purchased for very little and is particularly recommended for self-help builders doing their own plumbing or electrical work. It is valuable for cutting out pieces of framing that would otherwise have to be done with a hammer and chisel. There are those who believe that hand tools should be used exclusively for jobs such as these, that owner-builders should follow the advice of William Morris and leave the mark of the hand on their work whenever possible. I sympathize with this view, but my own inclination is for the amateur to reduce the amount of energy expended on unfamiliar tasks. Ease of accomplishment and the elimination of unnecessary frustration and stress make for more pleasureful work. Power tools get things done quickly and require little skill to master. If a builder already knows how to use a comparable hand tool with proficiency, if pride and pleasure are experienced in that ability, if time is not pressing—a lot of ifs—then by all means choose hand tools over power tools. Decisions such as this are the distinguishing mark of the terpitect.

A router was frequently mentioned by my sample of owner-builders as a desirable power tool. This is a specialized tool that is a virtual necessity for anyone doing finished carpentry. If a builder makes his or her own doors, windows, or cabinets, a router will produce the rabbet, dado, and dovetail cuts that are the benchmark of fine carpentry. However, I did not purchase a router; I found that my radial-arm saw was able to do most of the same work. A specialized tool that is an absolute necessity for log builders is a gasoline-powered chain saw. The best advice

I've heard from builders on purchasing this tool is buy the largest model you can afford. I purchased a twelve-inch chain saw to cut limbs and trees from my heavily forested lot, and I often wish I had selected a larger, heavy-duty model. The smaller saw will not handle large, green logs with any degree of efficiency. Nor will it produce large amounts of cut firewood.

Electric power drills are so inexpensive that a lightweight ¼-inch model is already in the toolbox of a majority of Americans. For house building, a heavy-duty model will be needed for some tasks, especially wiring, which can require an inordinate number of holes to be drilled through thick, dimensional lumber. It may be wiser to rent a special electrician's drill designed specifically for this purpose. Drilling through masonry requires a heavy-duty, ½-inch drill rather than the lightweight models.

A number of self-help builders in my sample listed power sanders as particularly useful tools, but I did not find this to be the case. There were times when I could have used a belt sander or a power plane—which remove large amounts of material quickly—but I did not have enough use for these tools to justify their purchase. I bought a lightweight pad sander and found that I could do everything it did much more effectively and accurately by hand. Among other marginal power tools which were mentioned in my sample as helpful were an air compressor and paint sprayer, a pneumatic nailer, and a jointer.

Ladders are essential pieces of equipment for house construction. Depending on which systems are self-built, several different kinds and sizes of ladders will be needed. To get on the roof, an extension ladder is necessary. A stepladder, which opens to form an A-frame, is used where there is nothing to lean a ladder against. The stepladder will be in constant use during all phases of construction; it is wise to get a large, sturdy one. A small stepladder, two- to four-feet high, will also be useful because it is easily moved around. Where a permanent ladder is needed before stairs are constructed, a ladder can be built from scrap materials. I made a ten-foot ladder from 2 × 4's left over from concrete forms, and it got me from the basement to the house platform for months. I also found that simple square boxes, two feet on a side and constructed from scrap lumber, make sturdy stools for reaching those places that are always just beyond arm's length.

Another indispensable fixture on a construction site is the saw horse. I ended up with a total of six before I finished my house, and all of them are now being used as the base for work tables of one sort or another. Saw horses are a good, basic carpentry task for a fledgling self-help builder; constructing them will acquaint the builder with the simple skills of measuring, cutting, and joining. The owner-builder can't have too many saw horses; they will quickly be put to use, not only for cutting lumber and plywood, but also for scaffolds, stools, and benches. Every carpenter seems to have an individual design and size for saw horses. Available scrap material often dictates the kind of saw horse that is built.

Construction equipment such as ladders, saw horses, and scaffolds are not the source of any joy in house building, however; they simply get you where the work is or hold it steady for you. Power tools, on the other hand, can magnify tremendously your own physical strength and can produce a sense of mastery over physical materials. I believe it is the *power* of power tools which is the source of much of the pleasure we derive from them, even more than their ability to get a task finished easily and quickly. Unlike hand tools, they are not easily taken for granted; you cannot ignore how powerfully they extend your human reach. They vibrate in your grasp, scream in your ear, and blur in your sight—a constant reminder of their energy and their threat. You cannot relax in their company; if you do, like one of those great, caged cats, they will devour your flesh.

Hand tools, however, are a true extension of the body. Their use involves much more than giving a mechanical advantage to the hand. The way a tool fits in your palm, its weight and balance, the texture of its handle in your grip—these are phenomenological facts that have little relationship to objective time-motion patterns or performance norms. But they do relate directly to how you feel about the work you are doing, so they are important considerations in determining the amount of satisfaction and pleasure you derive from your work.

Tools should be chosen, therefore, with care. Well-designed, well-made tools are invariably expensive, and telling a cost-conscious self-help builder that they will last a lifetime, as many of them will, is no help at all. The builder is mainly interested in their lasting through the construction of a house; durability beyond that is not a major consideration. Every do-it-yourself guide you read will advise you to purchase only the best tools; but this is sometimes poor advice, for there are many applications for which there is no justification for purchasing a high-quality tool. If a tool is to be used once or twice for a specialized task, there is little reason to invest a great deal of money in it unless collecting tools is something of a hobby.

For example, I purchased an offset hex wrench, a special plumber's tool for tightening nuts beneath sinks and lavatories. It is an essential tool for that one particular job, and a professional plumber probably uses it daily, but I am not likely to use it again for a long time, if ever. I bought one for two dollars, at the bargain table of a discount store, and it was exactly the right price for the right quality tool. I have purchased many tools this way and have never been disappointed. However, it is a mistake to skimp when purchasing an essential tool.

Chisels, saws, saw blades, drill bits, hammers, screwdrivers, pliers, wrenches, trowels, shovels, measuring tapes and rules: tools that are used constantly should be the best. The large mail-order houses—Sears and Ward—have established reputations for quality tools, as has Stanley. There are, however, many manufacturers of fine hand tools, and they can usually be identified by price. Well-machined tools of hard-tempered steel are expensive, and they usually show their pedigrees on close examination. If tools are purchased where professionals buy them, the amateur will usually not go amiss. Used hand tools can be a bargain if they can be found.

An owner-builder will go through a number of saw blades and drill bits before completing a house. Carbide-tipped blades are well worth the high price; they save the exasperation of trying to cut with a blade that is not yet worthless but not quite ready to be discarded. Because the builder will be drilling into hidden nails repeatedly, and one nail will wipe out a regular drill bit, high-speed drill bits for cutting through steel are the best kind for every purpose. Keeping cutting tools sharp is always a problem. Chisels, knives, and even the disposable blades of a utility knife can easily be hand sharpened on an emery stone, but badly knicked tools need a grinder. Saws and saw blades can also be hand filed, but most builders find it more convenient to take them to a hardware store to be professionally sharpened. Professional carpenters usually throw them away when they are dull because it is cheaper to replace them than to take the time and effort to resharpen them.

Tools are, of course, not an end in themselves, except perhaps to the collector, but a means of getting work done. Yet they are intimately involved in the process of building which can and should be a satisfaction in itself. "It is not the thing done or made which is beautiful, but the doing," Jacob Bronowski wrote. The creativity in constructing one's own house is not in contemplating or even living in the finished product, but in the act of building it. This is what distinguishes a terpitect from a contractor. Tools are a crucial link between worker and materials, for they make work possible. The ongoing joy of making one's own shelter, then, cannot be separated from the tools used in constructing it. There is an aesthetic pleasure in the tool itself; in its pathways over the work; in the rhythmic sequences of action dictated by its peculiar configurations; in the articulation of tool, arm, and hand over a familiar workspace. These ways of working are, to use David Sudnow's term, "the ways of the hand," as in improvising on the piano or in typing, where "a strict synchrony is sustained between any sayings you may be saying to yourself and the movements of the fingers" (*Ways of the Hand, The Organization of Improvised Conduct,* Cambridge: Harvard University Press, 1978, p. 90).

The shapes of basic hand tools have evolved over long periods of time, and their simplicity and elegance are refinements of small but significant improvements by anonymous craftsmen. The graceful curve of the ax handle, for example, has long been admired as one of the most efficient and beautiful evolutions of a hand tool in the history of human design. Before power saws, it enabled generations of wood cutters to work tirelessly from dawn to dusk, clearing forests and producing firewood.

A house builder who learns to appreciate the tools he or she works with not as simple implements of toil, but sources of gratification in work, will find in them a renewable pleasure.

Like craftsmen of earlier periods whose sharpened, polished tools passed from father to son as a personal legacy, warm friendships will be formed with hand tools. Clasping a favorite tool is a handshake with past labor and a link with what one is now building. In the course of finishing your house, tools and shelter become bound together through the intermediary of the builder's hand, so that tools become an ingredient of terpitecture.

SIX

HOME WORK:
THE PHYSIO-PSYCHO-
SOCIO-LOGICS OF LABOR

Those is something of a
paradox in the work that an owner-builder does on a house.
It is for the most part unfamiliar labor for which he or she has
no training or even skill; and therefore it is fraught with hesitation, uncertainty, and anxiety. Much of it is hard physical
toil which would never even be considered in any other context.
Yet this onerous, stressful, difficult labor is done of one's own
free will, in leisure time, without pay. It is open to question
whether an activity we do in our spare time, of our own volition,
at our own convenience—no matter how physically or emotionally exhausting—is work. Games such as tennis, golf, or
softball; sports like hunting, swimming, and running; chess,
cards, sewing—all of these activities are either physically, men-
tally, or emotionally exerting, but they are not "work."

If building a house is a hobby, like racing motorcycles or
philately, isn't it play rather than work? To answer this question
demands that some kind of distinction be drawn between work
and play. One major difference is that work is instrumental;
it has a product in sight, but play is an end in itself. In play

we attempt "to master the environment for the sheer pleasure of doing so," Walter S. Neff writes. "We struggle with the environment in work for compelling material reasons" (*Work and Human Behavior*, Chicago: Aldine Publishing Co., 1968, p. 82). According to this definition, building one's own house is work in the same way that playing tennis is work to the professional who does it for a living. The fact that self-help builders construct houses during leisure time does not alter the fact that they are creating shelter for themselves and their families, not merely engaging in a pastime for its own sake. The builder is, in effect, moonlighting on his or her own job, and it is every bit as much work as moonlighting for wages. The pay received is "sweat equity," an increase in the value of property equal to the market value of labor. If there is a mortgage, the builder receives this equity monthly in the form of reduced mortgage payments. If the house has been built for cash, sweat equity is recovered when the home is sold.

There is something wrong with the above definition, however. It implies that only play is pleasureful, that work must be utilitarian, distasteful, and unpleasant. Everyone is aware that this is simply not true; work can produce a wide range of positive feelings, from thrilling and exciting to benignly pleasant. The missing ingredient in our definition is creativity. Abraham Maslow defines creativity in such a way as to sound remarkably like play. The creative person, Maslow writes, treats work as "something *per se*, with its own right to be, rather than as a means to some end other than itself; i.e. as a tool for some extrinsic purpose" (*The Farther Reaches of Human Nature*, New York: Viking, 1971, p. 68). In creative work, then, work and play merge. Work becomes not merely instrumental; it also exists for its own sake. This is the kind of creative work a terpitect engages in—he or she builds for pleasure.

If building one's own house is to remain a creative experience, certain attitudes about ends and means must be established early in the enterprise and vigorously held to. Because an owner-builder is nearly always working under the pressures of time, there is a strong temptation to consider all work in terms of completion. Many house systems must be started before a previous system is completed; therefore, it is important to rush a job through to its finish. This means that the builder is always in a frenzy to finish the present task, in a frenzy to complete the next one, and so on. This is even more true if there are subcontractors involved who will arrive on schedule whether the preparatory work is done or not.

Working toward a deadline, or merely working with the finished product in view, is somewhat like preparing a meal for the purpose of eating it, and then wolfing it down to appease hunger. Nowhere in this scheme is there any creativity or joy, neither in the preparation of the meal, with its rich potential for skill and satisfaction, nor in eating it. Dining is reduced to consuming; it is not a process to be savored and enjoyed bite by bite, sip by sip; it is a finished fact. Between the start of preparation and a full belly there is only the vapid passing of food, like a computer throughput. At the other extreme is the Japanese tea ceremony, where each gesture of an elaborate ritual of food preparation is heightened and transformed into a moment of luminous import. The end product, tea, has little to do with the ceremony; a teabag can produce the same result. The act of ritualizing endows an event with significance; it has the power to transform water into wine, as it were, to convert work into art.

Constructing your own house can be a ritual act if you take care to allow the possibilities inherent in making a shelter to emerge. To a casual bystander, an owner-builder at work may appear to be no different from a carpenter or electrician working for wages. But the difference between the subjective experience of a terpitect and that of a paid construction worker is profound. In creating shelter, an owner-builder engages in an enterprise with psychological and cultural roots as deep as civilization itself. The work done on your own shelter is closer to the work of a musician, sculptor, or painter than to what it ostensibly resembles. It is self-actualizing and creative in the same way that traditional "high" art is, but not for the same reasons. The musician, sculptor, or architect is creative because a complex series of skills has been mastered and the artist is able to manipulate them in a novel and innovative way. The terpitect, on the other hand, is creative for precisely opposite reasons: he or she is an amateur whose skills are minimal, but who is constructing an artifact that will be judged by professional standards. Much of the creativity is derived from reinventing the wheel, discovering for oneself the craft wisdom of the ages. It is not innovation that bestows upon such work its creative powers, but standardization—accomplishing for yourself, often with painstaking and laborious effort, the same task that any competent craftsman could, and does, do thoughtlessly and tediously.

This kind of creativity is not public and negotiable, but private and retentive. The terpitect builds for self and family; he or she builds to last. The greatest creative thrill may come from merely duplicating a structure that was professionally built with apparently no more innovation than is seen in the repetitive cookie cutter models of a tract developer. Indeed, most owner-built homes cannot at a glance be distinguished from professionally constructed houses. If one defines creativity only in terms of product, then the terpitect is seldom a creator, for his or her house is often assertively ordinary, but if creativity is defined as a way of thinking and feeling, as it is in conceptual art, then the popular art of the owner-builder is as creative as the most prestigious high art.

Critics of owner-built housing argue that a self-help builder's duplication of what can be purchased makes no economic sense. They think that owner-builders would be far better off if they earned—at their own professions, where they are skilled and competent—the money to hire professionals. This is true only if the self-built house is viewed as an inferior alternative to a professionally constructed house, its only advantage being that it was built more cheaply. If a self-built house is considered one of life's most creative acts, however, then the economic argument becomes meaningless. The important fact is not how little of the work can I do myself, but how much. There is no doubt a degree of hubris in this—one can match oneself against the construction gods—but without pride there is very little creativity. There is also a certain amount of elitism among owner-builders who have constructed their homes without any professional help. In my sample of self-help builders I found a leitmotif: "I did it all myself, every board, every nail." It is something like the mystique of the professional athlete going the distance—all the way. I know something of this feeling, because even though I did not undertake all of the systems of my house myself, I built most of them. In the beginning I had little notion of constructing an entire house myself; it didn't seem within the realm of possibility. But as each new part of the house was completed, my confidence in my abilities to master each new challenge grew, until I was eager to finish as much of the house myself as was possible. The fact that I did not have the skills for each new task was of little concern, because I had already proved to myself that I could solve, to my satisfaction, the problems of inadequate knowledge and ability.

The work that one does on one's house has more to do with self-esteem, mastery, reputation, and prestige than with economics. Our modern industrial society offers few opportunities for us to satisfy these needs tangibly. Millions of Americans earn their livings at jobs that offer little self-satisfaction, merely a paycheck. These people vaguely reach out for ways to express their yearning for personal identity, for something that will leave a permanent mark upon a transient, anonymous life. They pursue hobbies and crafts, they purchase and collect, but remain unsatisfied. Most of the means they choose to give significance to their lives lack scale; they need a task difficult and arduous enough to challenge their strength, competence, and courage.

Building one's own dwelling is such a task, for a modern house, no matter what the size, is a formidable structure for an amateur to construct. Each of its numerous systems is a challenge for the person who approaches them in innocence and wonder. Consider plumbing, for example. Of course, we never actually *consider* plumbing; it is simply there, unless it goes wrong, and then we usually hire someone, at exorbitant cost, to make it right. But the person building a house must consider plumbing. It is not merely something that produces hot and cold water or flushes away waste. It is an essential subsystem that must be integrated into a number of other subsystems; it must be planned for, executed, tested, and maintained.

Designing a plumbing system—selecting bath and kitchen fixtures, connecting pipe joints, making hundreds of decisions and solving dozens of problems—these are the realities of the *work* of plumbing for the home builder. Even if he or she chooses to have a professional install the system, or only to rough it in, plumbing is much more than exasperation at a flushless toilet or despair at the cost of getting a leak fixed; it is a creative fact. The pleasure of turning a faucet for the first time and watching water flow from pipes you have installed yourself is not to be underestimated. The fact that plumbing is an invisible system, concealed behind walls and floors, detracts little from the feeling of accomplishment in learning its intricacies and solving its mysteries. And the knowledge that you are its creator and master, able to plug its leaks, replace its aging parts, unstop its arteries, and still its occasional whines—this knowledge is as comforting as knowing an honest car mechanic.

So it is with each of the subsystems of the house; the seemingly simple entities you have mindlessly lived with all of your life suddenly shove their way into your consciousness. Walls, ceilings, electrical outlets, stairs, closets, cabinets—who has ever given them a second thought? They are simply there; how they got there, what they are made of, never enters our minds. But when you have to design and build these structures which have surrounded you all your life and which were virtually invisible you *see* them for the first time. They are no longer passive architectural obstructions, but challenges to be met, creative possibilities. This is why someone who has never built his or her own house can never fully understand the intense pride that terpitects feel for the fruits of their labors. To the home buyer a wall is just a wall; to the owner-builder it is a hurdle bounded, a problem solved, a skill mastered, a new identity assumed.

This challenge of creating something of size and difficulty is subjective and individualistic; each of us has our own private sense of scale. The size of a house has little to do with the personal satisfaction derived from building it. There are perfectionists who lavish time and effort on acquiring and improving craftsmanship and on executing details. Others are concerned mostly with design ideas; the details of finishing are less important than architectural effect. There are some who get carried away with the sheer exuberance of building, who, like

Simon Rodia, the Italian-American mason who worked for forty years building the Watts Towers, fall in love with construction. Others, like the thirty-four-year-old mechanic who started by building an addition to his trailer, simply cannot stop. Six years later, working almost entirely alone, he completed a sixteen-room, three-story, colonial-style home. All traces of the mobile home have vanished; it is buried in the structure of the huge house. What remains is a populist pyramid, a contemporary monument to *homo architectus,* to a man's lust for building.

SELF-DISCIPLINE

Of course, there are the more mundane aspects of working on one's own house. Flights of creativity will come, but the depths of frustration, anger, tedium, and laziness are as frequent and must be dealt with. The work of building a house, like other kinds of work, involves organization, scheduling, and, above all, self-discipline. If one is able to manage these practical and functional parts of work, the creative raptures will take care of themselves. The pursuit of creativity, like the pursuit of happiness, is a myth; we go about our chosen work with love and care, and the translucent moments appear silently and unexpectedly, like fireflies at dusk.

One of the problems that almost all owner-builders must cope with is the excess of freedom in organizing work schedules. Those who have never been their own boss may find this freedom to be tyranny. They are faced with a project of enormous magnitude which possibly is going to take years to complete, and there is no one to insist that they show up on time, work a given number of hours, and not drink on the job. The habit of working must be established early and maintained; work schedules must be followed rigorously if progress is to continue. In the first flush of construction when enthusiasm runs high it is not difficult to forego the habitual pleasures of leisure time for hard manual labor, but as weeks roll into months and systems never seem to get finished, the self-help builder must rely on raw self-discipline to remain on the building site.

The one advantage the owner-builder has over a professional

worker who suffers from boredom or ennui on the job is that he or she can change to another task. There is no time during the construction of an owner-built home when a second or third system cannot be worked on to vary the tedium of a given job. There are usually dozens of unstarted, partially complete, or nearly complete tasks awaiting attention. By switching jobs for a few days, the builder returns to the original task with renewed energy and enthusiasm. In a contractor or developer house, each crew of subcontractors arrives sequentially, finishes its system and disappears, never to be seen again. The process has a mathematical regularity and is therefore the most efficient and economical way of building (but not necessarily the most human or the most creative). The owner-builder works much more organically, according to moods, whims, or personal prejudices. The more of the house systems you build alone, the more you are likely to build, for then you do not have to worry about scheduling subcontractors into your personal and idiosyncratic work plans. Autonomy makes this possible; you are your own supervisor, foreman, and laborer. If you awaken and decide, "Today, I do masonry," then masonry it is. If you feel carpenterish, you pick up hammer and saw instead of trowel. If there are helpers involved, this total flexibility is compromised somewhat, but not entirely, because you are the boss and are usually in a position to juggle the working times of your unskilled help.

In spite of the sequential systems approach which dictates certain tasks at certain times, the owner-builder is not like the chambered nautilus, the crustacean that labors on its current room and then seals each year's work behind it, never to see it again. The terpitect builds as coral does, by accretion, organically adding here and there, modifying, altering, rebuilding. Work is shaped to one's own psychological and biological rhythms, unlike the labor of the construction worker, who is a punchcard in a systems machine. This is a loose and open way of getting things done, and there is a danger that an owner-builder may not be able to manage it if he or she has never before experienced this much freedom.

The worker who most closely approximates this work style is not the single proprietor businessman, as one might expect;

he is usually tied inflexibly to his store or shop during the hours it is open. It is, instead, the farmer, who is bound only by the dictates of season and weather; within these natural confines he decides which kinds of labor he is to do and when. This may be one reason why farmers have traditionally been the most numerous and successful owner-builders. Their habitual work style is perfectly adapted to self-help building; they are experienced in managing a complex business system, disciplining themselves to physical labor, seeing tasks through to completion, working organically with the rhythms of nature. The core of the owner-builder movement in this country for generations has been rural America, where independence, self-reliance, and hard work are bedrock virtues. Insofar as an owner-builder is able to adopt all or a significant number of these qualities, the enterprise will prosper.

QUALITY

To speak of a work style or a work personality is to acknowledge that we are as different in the way we labor as we are in the way we think, feel, or relate to other people. With the exception of those who work for themselves, like farmers or small businessmen, we all conform to work patterns set for us by our employers. We must adjust our personal styles to the demands of the job and resolve whatever conflicts are involved between our way of working and the company's. How well we make the adjustment and internalize it as reasonable and acceptable dictates how happy we are at our job, or indeed whether we continue to work for that employer at all.

Owner-builders would seem to be spared this conflict between an outwardly endorsed work pattern and their own particular ways of doing things; they are, after all, their own boss and they make all the decisions about how work is to proceed. This is seldom the case, however, for they are constantly torn between how they think it should be done and the advice of others. This advice may come unsolicited from friends, neighbors, helpers, family, or spouse, or solicited from professional construction workers, building suppliers, fellow owner-builders, or self-help books and magazines. This counsel may not carry the weight of an employer's rules, but because self-help builders are amateurs doing work they are usually uncertain about, any suggestion that they are doing it "wrong," the slightest hint that another way is the "standard" or "professional" way can fill them with apprehension.

Nowhere is this more apparent than in the question of quality. When owner-builders are asked why they constructed a house themselves, one of the most frequent responses is, "I wanted a well-built house and I couldn't find one for sale." There are literally hundreds of thousands of Americans who are convinced, rightly or wrongly, that it is impossible to find a quality house built by a contractor or developer. "The houses built today are crap," was the way one self-help builder put it. Another told me that he and his wife looked at new houses for a year before they decided to build their own. "We didn't see anything that we felt was worth it. They were asking high prices for inferior construction," he said. "We came to the conclusion that the only way we could get the quality we wanted at a price we could afford was to build the house ourselves."

These are extreme statements of a more moderate view held by most self-help builders: "If you're going to go to the trouble of building a house yourself," a retired insurance salesman told me, "you may as well do it right." Doing it right for the owner-builder usually means overbuilding to the professional, adding more materials and labor than are necessary for sound construction. What are adequate and acceptable work methods to the professional are often "short cuts" to the owner-builder who would rather take the time and the extra effort to build what he or she perceives to be a more durable, safe, or aesthetically pleasing structure.

A typical instance of this conflict between professional opinion and personal perception of building practices occurred when I was about to pour a concrete footing for my chimney. I had consulted a building manual and learned that the footing should be reinforced with steel because of the great weight it must support. When I went to the building supplier to purchase the steel reinforcing bars I got into a conversation with a concrete contractor. "Ain't nobody puts steel in chimney footings around here," he opined. "Don't need 'em." Well, there I was, faced with conflicting expertise. One opinion carried the authority of the printed page; the other was based on local building practices. I would have preferred not to have to put the steel in; it would have saved me the cost of the material and the effort of cutting, bending, and tying the steel rebars, and my work style is to minimize labor on structures that will never be seen. After much hesitation, I did buy the steel and put it in the footing, feeling a little silly and very much an amateur as I did it. But when I poured the concrete into the footing forms and watched it ooze around those bars I felt a rush of pleasure—I knew I had made the proper choice; I was doing it the "right" way. I really had not given those steel bars another thought until I wrote this passage, but somewhere in the back of my consciousness they support, not only my chimney, but my own good opinion of myself. I had not taken a "short cut"; I had built with strength to spare. I think that most owner-builders would understand and approve of my decision.

After getting conflicting advice on whether or not to reinforce with steel the concrete footing for my chimney, I opted for the steel, a decision I felt was right when it was finished. A small concrete mixer is perfect for jobs like this.

CONFLICTS BETWEEN JOB AND HOUSE

Another conflict that often occurs during construction of a house is the opposing demands of avocation and vocation. Although most of the construction on a self-built house occurs during blocks of time released from a full-time job, it is inevitable that work also proceeds, often for long periods of time, during the free hours after a full day's work and on weekends. There are critics of the owner-builder phenomenon who argue, with some justification, that house builders are in reality stealing time from their employers, because the focus of interest, energy, and enthusiasm lies not with job but with house.

How much your occupation suffers while you are building a house will obviously vary with the individual. Those who do physical labor for a living may find that the strenuous work done after hours or during weekends will leave them stale and exhausted for their regular work. I believe this will occur only if the builder is doing labor that is not experienced as creative or self-fulfilling. Fatigue from self-actualizing work is quickly dissipated, and there is no reason why the emotional exhilaration of the building experience cannot carry over into one's vocational work. Those who do sedentary or intellectual work for a living may find that once they are in physical condition, the labor of house building is a source of renewed energy and stimulation for their intellectual work.

Builders whose vocational work is not fixed by office or shop hours, but who work irregularly at home as well, may have more difficulty in separating vocation from avocation. Building a house consumes large amounts of time spent planning, researching, pondering, worrying, reading, discussing, and even dreaming about the next day's or week's work. "Sometimes I lie awake for hours planning what I'll do tomorrow," one owner-builder told me. I know this experience well, for I've done it many times. The toll all of this takes on a regular job depends on the ability to throw a mental switch back and forth between vocational work and house work. To be able to close the door on the house when you leave the building site is a talent to be nurtured. Those who carry the cares of their job home with them, who find it impossible to relax because of office or plant worries, may not be able to carry the responsibilities of simultaneous and often conflicting work schedules comfortably. It is important for owner-builders to remember that their first allegiance is to their vocations. They may often feel after a particularly good day's work on the house that this wouldn't be a bad way to make a living, building houses. But the moment a terpitect becomes a contractor he or she steps into another world where profit rather than pleasure is the goal. Work is then of an altogether different kind; he or she is no longer creating shelter, but making a living, and the two march to different drummers.

AVOIDING ACCIDENTS

In discussing work styles I have omitted an important consideration: physical condition. Those who merely act as their own contractor and subcontract the rest of the house systems need not concern themselves with whether they are physically able to see the project through to completion. Most owner-builders, however, complete many of the house systems themselves, and therefore their state of health does matter. One would think that where health and safety are concerned, particularly in an enterprise that is so determinedly physical, age would be a critical factor, but I was surprised to learn that this is not necessarily the case. When I began my interviews with owner-builders I was not prepared for the number of men and women in their middle and advanced years who built their own houses. I had thought that it was most certainly an avocation for the young, but this is not so; there are thousands of couples who become terpitects after their retirement. Others who have already completed owner-built homes plan to build retirement homes in the future. The mean age of those I questioned, the age at which they built their houses, was thirty-nine, but this included teen-agers and people in their seventies.

One of the questions I put to owner-builders was, "How hazardous is house building to your health; can you get hurt?"

The responses were about evenly divided between those who said that with reasonable care it was not particularly hazardous and those who affirmed that it was indeed intrinsically dangerous. Those who injured themselves seriously made up a very small minority; there were broken bones, usually from roofs, scaffolds, or ladders; cuts and lacerations; and various injuries to back and joints. My own opinion is that for the amateur, much more so than for the professional, building a house can be dangerous; and one must be constantly on the alert to avoid injuries. On the other hand, it is no more hazardous than most contact sports, and the minor injuries are about the same as those one must inevitably suffer from such sports as tennis, running, swimming or diving, gymnastics, or weight training.

Physical conditioning is, of course, the key to avoiding injuries to muscles and joints, and constant care is the only way to prevent serious accidents. The amount of physical exertion involved in building the systems of a house range from the mildly tiring exercise of painting to the hard labor of concrete and masonry work. Every would-be terpitect should be aware, however, that even though he or she may have occasionally fastened lumber with hammer and nails, wielding a heavy framing hammer for hours on end is physically exhausting and potentially damaging work for anyone who is out of condition. Before choosing to construct any of the more arduous house systems, every owner-builder should take stock of his or her body resources. If there is trouble with knees or back, for example, concrete and masonry work should probably be avoided. The dust from finishing sheet rock, the fine particles of glass from insulation, as well as sawdust from carpentry, may all be inimical to those with respiratory or allergy difficulties. Anyone who is seriously overweight or has a heart condition should obviously have a medical examination before doing strenuous physical labor.

In construction, as in athletics, fatigue usually leads to accidents. One builder, for example, lost the tip of his left index finger when he ran it over a jointer. "It was late and I was tired and got careless," he reported. By taking a few simple precautions, however, anyone in reasonably good health should be

More owner-builder accidents happen on ladders than on any other piece of equipment. The worker on this urban homestead project is stretching laterally, making it dangerously likely that the ladder will tip over. (Photo: HUD)

able to complete all of the systems of a house unscathed. The following safety suggestions were offered by a variety of owner-builders:

Give ladders and power saws particular respect. You are unusually vulnerable when using them. Ladders and scaffolds put one within bone-breaking distance from the earth, and power saws are unforgiving of the smallest mistake. As one owner-builder observed: "Although I've been using my radial-arm saw for a number of years, it still scares the hell out of me—and I want to keep it that way."

Those who design steep roofs into their houses should recognize that they are creating a permanent hazard.

Any time two people are working on a job, the one working below the other should wear a hard hat—even in hot weather when the hats are miserably uncomfortable. One owner-builder told me of dropping a brick from a scaffold onto his wife standing below. "Fortunately, she wasn't seriously hurt," he reported, "but it was hard hats after that."

Wear a pair of workers' shoes; a nail through the foot is one of the most common injuries. One of Parkinson's minor laws states that hammers always gravitate to fingers and falling objects to feet.

Lift heavy objects with leg muscles instead of back muscles; a low backache puts the builder out of business for weeks or months.

Always wear safety glasses when doing anything that can possibly injure the eyes. Like hard hats, safety glasses are a nuisance; they fog up and get scratched, but eyes are *very* special.

Beer and hard work complement each other like—well, like all the ways those television commercials show us. Getting a

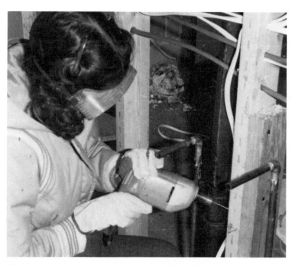

Wearing protective glasses when using power tools is a nuisance, and for making routine cuts on lumber they are usually unnecessary. But whenever a power tool is used on metal, masonry, plastic, or wood that may splinter, they should always be worn. They are sweaty and uncomfortable, but are definitely worth the inconvenience.

little high on a couple of beers *after* a hard day's work and surveying the new addition to the house is one of the pleasures that no terpitect should deny himself or herself. But drinking while working—something that professionals are never allowed to do, but which owner-builders, because they are their own boss, can do any time they want to—can lead to careless mistakes and bad accidents.

The worst injuries sometimes come from jobs or substances not considered harmful. A North Carolina owner-builder, for example, seriously burned his feet while pouring concrete; wet concrete got over the top and inside his boots. This kind of accident is easy to fall prey to because wet concrete is not considered a harmful material, but it is highly caustic when pressed next to the skin for any length of time.

One owner-builder offered this final piece of advice: "Anytime you take on a new job always start by asking yourself, how can this hurt me? Then take the necessary precautions."

PSYCHOLOGICAL PARAMETERS OF SELF-HELP BUILDING

Even more than physical condition, an owner-builder's emotional state can affect decisions about work. Often a self-help builder has the desire to build a house system but is afraid to try, for a number of reasons. The anxieties associated with making such choices are not synthetic; they can be as real as a whiplash. These stresses may center on whether there is enough time to complete the system or whether the task is simply too big and complicated to do oneself. Such anxieties cannot be dismissed lightly. The decision to lay brick on the exterior of a house or even to build a masonry fireplace, for example, is a major commitment in time and work. It may mean handling literally thousands of bricks; lifting bricks and mortar up on a scaffold and laying them one by one. Such a job may seem truly overwhelming, and the very contemplation of doing it may provoke severe anxiety.

There are numerous occasions during the construction of a house when self-help builders may suffer such a crisis of nerve.

At these times they will need to remind themselves that all large enterprises are accomplished one day at a time; complex systems can be reduced to webs of simple subsystems; and weighty problems often succumb to hairspring solutions. Nevertheless, there will be times when a negative decision is the only sound one; working on a house system which will consume unacceptable reserves of emotional energy will invariably lead to pain not pleasure, and the thesis of this book is that building one's own house should above all be a richly satisfying experience. Not all owner-builders are successful in meeting the psychological demands of building house systems themselves. There were several I encountered who declared that if they had it to do over again they would subcontract the entire house. In most of these cases, the owner-builder felt that the money saved was not compensation enough for the work—physical and emotional—involved.

In such instances, decisions about work were derived from economic rather than personal considerations. In *The Farther Reaches of Human Nature*, Abraham Maslow speaks of the idea of a "calling" which breaks down the polarity between work and pleasure. The calling to which an owner-builder responds is one of the most primitive of all, the desire to make one's own shelter. A calling has traditionally been directed toward spiritual yearnings, but it is well to remember that when St. Francis received his call, he turned to rebuilding, with his own hands, the abandoned chapels in the vicinity of Assisi. Building a house for God is a calling that drove millions of men and women to construct the great cathedrals of medieval Europe, and it is not too far removed from the calling of building a house for oneself.

One final statement ought to be made about how an owner-builder internalizes work on a house and, in doing so, transforms toil into terpitecture. In this chapter I have related house building to theories of need gratification and self-actualization, to the fusion of work and play. I have argued, in effect, that building one's own house is an emotionally healthy act. Just as self-actualized people are without neurotic or psychopathic inclinations, according to Maslow, the man or woman who successfully builds a house has taken a resolute step in the direction of emotional well-being. I would certainly not assert that there are no neurotic owner-builders or that all of them are self-actualized individuals who have developed their talents to the fullest possible degree with freedom and joy. But insofar as building your own house is a creative, self-actualizing experience, it is psychologically enriching and healthful, and living in an owner-built home is a source of continuing personal gratification.

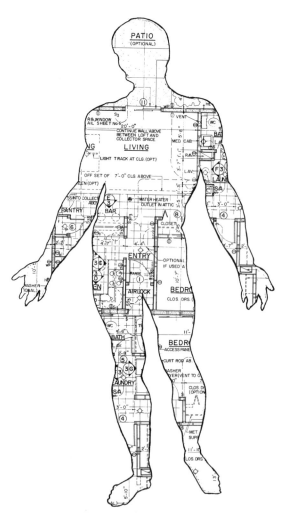

Owner-builders re-create themselves in their homes.

SEVEN

HURRY UP—OUCH!
TIME AND ERROR

ne of the eco-
nomic facts of the self-built home is that time—like money,
materials, and labor—is a scarce commodity, accountable to the
laws of supply and demand. The less there is of it, the more
valuable it is. Because the owner-builder is involved in an
avocation, the time expended on it is borrowed from other
sources: family, friendships, leisure pursuits, sports, or a second
job. Time is therefore almost always limited and scarce, and
must be used economically.

Like the professional builder, the self-help builder is a captive
of the seasons in the allocation of time. Spring, summer, and
fall are good building times; winter is terrible. Dry weather
allows work; wet weather often closes it down. Unfortunately,
the released time from a job does not always correspond to the
optimal seasons for building. For a farmer, for example, spring,
summer, and fall are the seasons that demand the most vocational time; winter is relatively free. Teachers have the summer
off—one reason why a large number of owner-builders are
teachers—and are able to build during the optimum season for
it. But they have little spare time and their weekends are not

139

always free during the rest of the year, because much of their work, such as grading and preparation, is done at home. Construction workers account for a large percentage of self-help builders because they usually possess some of the necessary skills, but also because they seldom work a full year. This means they have periods of time to devote to their house project, usually during winter months. Strikes, production changeovers, and layoffs offer self-help builders opportunities for blocks of work time, but only if planning is done in advance because the work stoppage often comes suddenly and unpredictably. Income, unfortunately, also stops when full-time work comes to a halt. A West Virginia coal miner and his family completed two houses during two protracted strikes three years apart. The sale of one of the houses helped them ride out the long periods of unemployment.

Different house systems make different kinds of demands for time. For instance, grading, footings, and concrete work require dry weather. Framing is not as critical, but once the builder begins to frame a house there is intense pressure to get it dried in. The sight of rain pouring down on an unroofed frame and plywood floors is a disconcerting sight for any owner-builder. The primary goal during this stage of construction is to get the rafters covered with roof decking and to get the roofing material in place. To stand under a newly-finished roof and watch rain pour off the eaves instead of through the rafters produces a warm, dry glow. Once the frame is closed with roofing and wall sheathing, the builder is released from the pressures of time. He or she can relax, work more slowly and deliberately, content in the knowledge that the weather is no longer an enemy.

Unlike the professional, the self-help builder is unlikely to work at a fixed pace. Work will be done in spurts, depending on the exigencies of the system being worked on and the state of one's own psychic economy. Once the roof is up, the builder is free to progress at a more leisurely pace. There will be other urgent times, but between them an owner-builder may wish to take a break from construction for a while and accumulate either money or energy. Time will enforce new demands when a system must be finished in preparation for the arrival of a subcontractor or when a date has been set for moving in. A large number of self-help builders move into their houses long before they are finished, if codes allow it. When a builder has decided to move into an incomplete house there is immediate pressure to complete the bare necessities: plumbing, electricity, kitchen, bath.

After moving into an unfinished house the time frame in which work is perceived may be sharply altered. A family is suddenly thrown into living conditions that would be unacceptable under any other circumstances, but are tolerated because they are known to be temporary. There may be limited privacy, little in the way of amenities, and few creature comforts. Dust, dirt, and litter are ubiquitous. Everyday tasks like cooking, clothes washing, and personal hygiene are reduced to primitive makeshift operations. Yet adaptations are made quickly because builder and family accept the inconveniences and disruptions of habitual life style as the kinds of temporary disorders that occur when a major move is made, or during vacations, or camping trips. Indeed, the analogy to camping is sometimes so close that the term "camping in" is often used to describe these hectic times.

Once a family is living in an unfinished house the progress of construction is likely to change from the regular sequences of a professionally built house to an organic growth based on the family's continuing adjustments to its new environment. Inhabiting a space significantly alters perceptions of it. The space expands or diminishes, becomes public or private, cozy or impersonal, comfortable or uncomfortable, depending on a subjective gestalt of spatial configuration, personal distancing, temperature, light, and materials. House systems will be modified, scrapped, or added to; space will be altered; and such aesthetic considerations as textures and colors will be changed because of the experience of living in what was previously only a conceptual space.

Time schedules and work priorities will also shift; work will be concentrated on those systems where the need for completion is most urgent—kitchen and bathrooms, for instance—even though this work may occur out of the normal sequence dictated by standard building practices. This often results in redundant work, inefficiency, and frustration, but this is the price that must be paid for substituting a personal, organic style for efficient work. When you haven't been able to take a shower in weeks, bathroom plumbing receives top priority, even if it means that later jobs will be harder to do.

It is during feverish work periods, when a self-imposed deadline has made time scarce, that the self-help builder becomes aware of how difficult it is to judge the time any house system is going to take to finish if it is self-built. It is inevitable that when plans are made to construct a system, not only is a price tag put on it, but also a time tag. "I ought to be able to finish this job in three weeks," a builder will say. This estimate is a ballpark guess based on how he or she has worked in the past, how many hours a day can be given to a job, how much help there is, the availability of materials, of good weather, good labor, and good luck. It is fair to say that this estimate will, nine times out of ten, be absolutely wrong. It will not take three weeks; it will much more likely take six weeks—probably at least nine.

Owner-builders are inveterate optimists about how much time a task will take. This is understandable, for no one but an optimist would undertake building a house in the first place. This optimism takes the form of a number of untested assumptions, the shakiest being that you will be able to quickly master a skill you know virtually nothing about. Learning carpentry, masonry, plumbing, or electrical work, as I have already indicated, is not particularly difficult, but it is slow. The main difference between the way a professional does a job and the way an amateur does it is not always the quality of the work but the time it takes. All the tricks of the trade a professional construction worker masters—and they are numerous—do not usually produce better work but faster work. An amateur can duplicate the level of quality of any construction worker if he or she has the tools and is willing to take the time. The greatest inefficiency is evident the first time any job is done, no matter how simple it is. If you can recall how long it took to change a car tire the first time you tried and then note how long it takes a garage mechanic to change one, you will get some idea of the time differential between amateur and professional doing the same, simple task.

Not all house systems are equally inefficient in the use of the owner-builder's time. Amateurs, for example, can paint or stain perhaps half as fast as a professional painter, but until they have laid thousands of bricks they are unlikely to be able to do in a week what a professional mason can do in a day. One reason for the time differentials between professional and amateur is that, although all of the building trades are labor intensive, some involve skills, tools, and technologies that are either difficult to master, costly, or unavailable to the amateur. The most obvious example is grading, a system that requires heavy machinery. It may be possible for builders to rent or borrow a bulldozer, but unless they have the skill to use it or wish to acquire the skill and are willing to pay the costs, they are not likely to save time or money, or to experience a great deal of personal satisfaction in grading the site themselves.

In the case of interior painting, however, the opposite is the case. It is true that a painting crew, using compressors and spray guns, can paint the interior walls of a house in a single day; but the owner-builder, using a roller, can do the same job in two or three days. Little time is lost in acquiring the skill. Painting millwork and trim is hand work for both professional and amateur, and here the time advantage of the specialist is narrowed even more.

There is another reason why the owner-builder may work slowly compared to the professional—he or she is enjoying the work so much that there is no hurry to complete it. This can set up a conflict because we have been taught that the only way to labor is efficiently. Quicker is always better. But now, this basic truth is questioned. You may find yourself enjoying stretching out a job that could be done in a shorter amount of time by settling for less than perfection. But you choose to be inefficient, to overbuild, to take care even where the work will be covered up, like the Greek artisans who sculpted the backs of temple friezes even though they could never be seen from the ground. Or you choose to be playful or whimsical, and add an unplanned, outrageously time-consuming piece of work simply because it strikes your fancy. Sooner or later you will be forced to recognize that sometimes the important thing about building your own house is not getting it done, but doing it.

The difference is between means and ends, process and product. Having finished a house to live in becomes less important than building your house. I think this is often not consciously recognized, and accounts in some cases for building that goes on and on, nothing ever getting completely finished for fear of ending the pleasures of constructing. Mixed feelings of satisfaction and regret often accompany the completion of house systems. Cases like this illustrate forcefully how building a house satisfies personal needs for creativity and self-fulfillment. It is a further instance of builders re-creating themselves in their shelter.

CRAFT MEMORIES

There is yet another way that time steals into a terpitect's enterprise, less obviously but no less deeply felt. We all live in a world where the present is, to use William James's term, "a rainbow over the waterfall"—a moment of permanence through which the future flows down into the past. We conceptualize a linear present which is a series of discrete pinpoints, but we live in an experiential present that is "extended." There is a fringe which anticipates the future and another that fades into the past.

Into this extended present we draw the past as memory. This is the dynamic process by which the past is renewed and revitalized in our consciousness, and remains a living part of our present rather than a fossil in the past. In the case of the creative act, this process is usually connected with an artifact: a painting, a piece of sculpture, a tune, a poem, a novel, an invention—a house. Embedded in this object are "creative memories" or "craft memories" which have significance only for the creator. The painter sees brush strokes in the finished painting which are reminders of the pleasures, anxieties, inspirations, of particular choices: the solution to the problem of the tilt of a lip in a portrait, the delicate shading of a shadow that defied the palette for hours. The composer recalls a twist of melody that brought a tingling to the fingers on the piano.

These creative memories live, in varying degrees, in every artistic creation. They can be known only to the artist or craftsman who made the object: they are qualitatively different from the memories of any other observer. They are what the dancer calls "body memories" and the actor "emotional memory"—purely subjective states attached to solutions of physical or psychological problems.

Shelter is a special category for the collection of craft memories, for it represents not a series of discrete works which may or may not remain in the possession of the creator. It is lived in, as a spatial environment; it engulfs us as the womb encloses the fetus.

You sit before your fireplace and watch the play of flames, as millions of homeowners do each winter; but your eye drifts upward to the stones that shape the hearth, and you recall the act of fitting individual stones into place and the emotions, both pleasurable and painful, that accompanied the act. You recall perhaps where you found particular rocks and how you fit them into the patterned mosaic which puzzles its way toward the ceiling. You note the rock you wished you had placed differently and those that particularly please you, and from this integration of past memory and present experience you produce an extended present that includes in its matrix a richness and density of living that no purchased shelter can supply.

Consider what part of your total experiential awareness is spent "at home." A vast portion of our lives is played out in the narrow confines of our home space. Of course, it is true that our field of vision filters out the surroundings most of the time; we focus on whatever interest consumes us at the moment. But it is during the periods of relaxation, meditation, inner quiet that we associate with the pleasure of hearth and home—our most deeply felt pleasure—it is during these moments that our creative memories are pulled into our extended present and infuse our pleasure with a kind of Wordsworthian "recollection in tranquility." Indeed, the home builder is storing up, as he or she shapes a house, pleasures that will be "felt in the blood, and felt along the heart" in years to come.

These creative memories can involve even structural elements hidden from sight. In my own house there is an elaborately carpentered piece of trusswork behind a panel in my living room. It serves no structural purpose, but this trusswork continues to give me great satisfaction.

A house still being framed is a precarious structure. Until it gets its board, plywood, or brick skin it is frighteningly unstable; that is why intricate series of braces are used to prop up walls during the framing process. It is possible to place your shoulder against the corner of the frame and watch the whole house sway drunkenly. To an amateur home builder this is an unsettling sight. He or she has read or has been told that the structure will achieve stability as corner bracing and more of its skins are applied, but nevertheless, there it is, rocking like a hobby horse.

The south side of my house is made almost entirely of glass, and for months this part of the building was no more than a series of slender posts supporting roof beams, tied together with what seemed to me fragile window frames. The whole thing would teeter at a touch. Between the clerestory and the lower windows was about a foot of wall that would support the weight of all that glass. I became convinced that it was totally inadequate, that glass would ultimately come crashing about me, so I made of that foot-wide space a sturdy truss that took me a day to construct.

I placed the plywood exterior over my truss and realized how silly my fears had been; once it was covered with plywood, that foot-wide wall was as strong and rigid as a steel beam, and my fancy trusswork was superfluous. Still, I admired it frequently while building the house and only with great reluctance did I finally cover it with interior paneling.

The trusswork remains there, out of sight but very much a craft memory. I still take pleasure in it, with the full knowledge that it is redundant, perhaps *because* it is redundant. Its existence is a permanent record of my insecurity and of my attempt to deal with it in a positive, creative way. The pleasure I had designing it, my satisfaction in executing it—these will be a part of the house as long as I live in it, and also a part of the pleasure the house affords me.

THE INEVITABILITY OF ERROR

It would be unfair to catalog the sources of satisfaction and pleasure in working on your own house without noting that there are those unhappy times when everything seems to go wrong. How these occasions are dealt with, so that they do not sour the pleasures of construction work, is as important in its own way as learning specific skills. When one has botched a job repeatedly, often a simple task, it is possible to develop frustrations, anger, and even self-loathing that saturates and contaminates the entire work day. Overcoming these feelings is a matter of developing new sets of attitudes and expectations about work.

All owner-builders embark on a personal voyage of discovery; in charting the unknown regions of their potential for creativity and self-fulfillment they must master the fear of inadequacy and failure. These take the form of a conviction that the job demands too much skill, that a series of major blunders will ensue, and the house will be structurally unsound or, even worse, will look like hell. All self-help builders suffer, in varying degrees, from errataphobia, the fear of making mistakes. It is a well-founded fear, for there are few owner-builders who can honestly state that they finished their houses without mistakes. In spite of the innumerable errors that self-help builders have recounted to me and the literally hundreds of mistakes that I have made myself, I am willing to go on record as stating that I don't believe owner-builders make many more errors than professionals do. And there is absolutely no debating the fact that they correct many more of the errors they make than professionals do. Every time a bridge collapses, a building sags, a window pops out from a skyscraper, or an electrical fire burns a building down, some professional has made a serious mistake that has not been fixed. Every community in the nation has its horror stories of unsound, unsafe, even criminal building practices that have resulted in the loss of life and property— all the result of professional mistakes. Anyone who has worked in the construction industry can tell endless tales of building errors that are quickly concealed and never reported. Simply take a walk around a developer housing tract during the framing stage of construction and you will see work that no owner-builder would ever allow to remain uncorrected. Owner-builders have no need to feel inferior to professional construction workers where mistakes are concerned. The owner-builder's mistakes come from not knowing; the professional's from not caring.

Mistakes will be made by the amateur, again and again and again. He or she soon learns a single cardinal rule, however: accept error as a necessary part of being an amateur and correct it as quickly as possible. The lost time and wasted materials are simply part of the costs of building one's own house. I still explode in anger at myself every time I measure a piece of lumber six inches too short. But the explosion is cathartic and I cut another board to the correct length, although I have on a number of occasions repeated the same error several times. Every self-help builder will recognize this experience. Once error is accepted as a necessary and inevitable ingredient of the owner-builder enterprise, one reacts to it philosophically: you call the board, the tool, or the tape measure a bastard; you call yourself a numbskull, and you go on with your work.

Those who have the most difficulty accepting the inevitability of error are those with compulsive personalities, who demand that life be precise and perfect. If you are uncomfortable unless everything is in its place, unless things are orderly and clean and done "the correct way,"mistakes can be a serious problem. If, in order to avoid error, an owner-builder takes too much time and care with work, nothing will ever get finished; work will drag on for interminable periods of time.

It is possible to avoid some of the more common errors by developing work habits that minimize human fallibility. Perhaps the most frequent mistakes are errors of measurement. These occur everywhere that materials must be cut and fitted, most commonly in carpentry, plumbing, and electrical work. They are less common in masonry and roofing, where the dimensions of materials are fixed. Mistakes in measuring materials are usually mental lapses rather than physical errors in marking or cutting. A good number of these result from working alone rather than in teams as professionals invariably do. For example, workers installing lapboard siding will usually have one person cutting boards while one or two are on ladders or a scaffold putting up the pieces. Measurements are taken and called down to the cutter, who instantly puts saw to lumber. An owner-builder working alone, however, must take the measurement, commit it to memory, climb down to the sawing location, then mark and cut the material. Between measurement and cut, enough time has elapsed for memory slips to occur, especially if there is an interruption of any kind or even if a stray thought enters the mind for an instant. The obvious way to prevent this is to write down the measurement, but there seems to be a psychological block against doing this for many builders: "I can surely remember a simple number for a minute or two; after all, I'm not an idiot." So pride goeth before a fall in one's self-esteem: the board is miscut and you feel like an idiot. A pad and pencil thrust into a carpenter's apron prevents most errors of this kind. But I must confess that even when the paper and pencil were handy, I often obstinately refused to write the measurement down, convinced it wasn't necessary. It was.

Other mistakes in measuring occur simply because of failure to observe the carpenter's rule: measure, mark it, check it, cut it. Professional carpenters, from long practice, read the markings on a rule as instinctively as we read print on a page, but it is very easy for an amateur carpenter to make the kinds of errors children make in reading—though for thought, or fight for flight—and read ⅝ for ⅜. Adding fractions also accounts for numerous errors. Constant care and checking are the only way to keep a rein on these kinds of mistakes; they are never likely to be completely eliminated.

Another kind of error occurs when small, seemingly insignificant lapses accumulate into a major blunder. This can easily happen in masonry work. Mortar joints must be held to precise size if the exactness of rows of fixed-dimension bricks is to be maintained. An accumulation of small errors suddenly presents the amateur mason with a space too small for the final brick in a row. The same kind of problem can occur if the spaces between studs are not accurately maintained. This error does not show up until much later when you install sheets of siding or sheet rock and suddenly find they don't meet over a stud.

An owner-builder will correct this kind of mistake by either cutting more sheet material or installing an extra stud, but professionals often nail up the siding or sheet rock with nothing behind the joint. Several years later this will assert itself in the form of warped siding or bulging walls.

Errors in design or omission are often much more serious and more difficult to fix. I inquired of my sample of owner-builders what their worst mistakes were and what they did about them. The responses ranged in seriousness from a design error that required supporting jacks to be placed under the floor to prevent it from sagging to failure to align the edges of wallpaper properly. Many of the mistakes were design errors incorporated into the project at the blueprint stage but not recognized until the house was nearly completed. Some wished they had made rooms either larger or smaller. Others wanted more windows or doors in different locations. Many design regrets had to do with energy features and were the result of the dramatic increase in energy costs in recent years. Some thought their houses were now too large and costly to heat, others regretted not installing central heat, a heat pump, a wood stove. Inadequate insulation is virtually a universal complaint. These are all hindsight errors, however; at the time of construction the energy shortage could not have been anticipated. Nevertheless, they illustrate the crucial importance of planning in order to avoid disappointment.

Two design errors that showed up repeatedly were the failure to include a basement and building with too small a crawl space. A basement represents extremely cheap space, and those who forgo the opportunity of utilizing this space sometimes regret it later when possessions threaten to burst the very seams of closets and storage rooms. Inadequate crawl space usually results from aesthetic considerations. It is easy to design a crawl space that is less than three feet high, the distance usually considered the minimum, because most one-story houses look better when they are lower to the ground. Ranch houses on slabs, with their long, low profile, are understandably one of the most popular designs in the nation. When the crawl space is too low, maintaining heating, plumbing, and electrical systems beneath the house usually means taking the word "crawl" literally—bellying through dirt and working on your back. It is the kind of error that grows in importance as the years progress.

Permanent structural errors were few, and where they existed and could not be repaired, the builders usually accepted them philosophically. As an Oregon woman who constructed most of the systems of her home herself put it: "Any mistakes I made I just put down to experience and figured I could live with them. They weren't any worse than some professionals make— and I didn't have to pay $15 an hour for them."

EIGHT

HELPING OUT:
SUBS, FLUBS,
AND NETWORKS

Few owner-builders finish their houses without any assistance at all. Most get help from a variety of sources: they receive free advice and labor from friends, business associates, family, and other owner-builders; or they hire skilled or unskilled construction workers. Managing this labor is one of the most difficult tasks for an amateur who is inexperienced in supervising construction work. It accounts for one of the greatest sources of anxiety in house building, and leads many self-help builders to conclude that it is often easier and less frustrating to do a job themselves rather than endure the worries and frustrations of supervising hired labor.

DIFFICULTIES WITH SUBCONTRACTORS

Subcontractors are by far the peskiest splinter in an owner-builder's handiwork. These are ordinarily private businessmen who manage teams of masons, carpenters, electricians, or plumbers. Sometimes there is a single proprietor—an electrician, say, who wires a house himself. In a contractor-built house the contractor manages the subcontractors; he or she hires them, sees that they arrive on schedule, and that the work contracted for is satisfactorily completed. Owner-builders who choose to perform only the planning/management function of the house take on these responsibilities themselves.

An owner-builder is at a distinct disadvantage in dealing with subcontractors; once they learn they are working for an amateur, with no opportunity for future jobs, they are more likely to be late, careless, or incompetent than if they were working steadily for a contractor. Subcontractors can be frustratingly undependable. They may promise anything to get the job, but fail to arrive on the expected date because a larger, more profitable contract turned up. Many of the delays experienced by subcontractors are unavoidable because of the sequential nature of the construction industry. Any holdup in one building sequence will cause all of the subsequent systems to fall behind schedule. Weather accounts for many of the delays; materials shortages, strikes, walkouts, labor shortages, and inefficiency can cause others. Subcontractors will seldom pull crew and equipment from one job to another unless it is completed; therefore, any delay in the current job will mean that the self-help builder must patiently—or impatiently—wait.

Anyone who hires a subcontractor to do work assumes that he is a responsible businessman, that he will fulfill his end of the contract, and complete the work satisfactorily. This may be an unwarranted assumption; there are some marginal operators who are not competent businessmen, who do not themselves know how to hire and manage labor, and who cannot be certain that their own workers will be sober, competent, or dependable. Talking about subcontractors with owner-builders is sometimes like discussing Satan with a fundamentalist—the language runs to hellfire and damnation. When I asked my sample of owner-builders what was the most difficult part of constructing a house, nearly 20 percent responded that it was dealing with subcontractors. Besides getting them to show up when promised, the complaints cited poor quality work, cheap materials, theft, drunkenness, and failure to complete contracted work.

Owner-builders offered various kinds of advice on how to deal with subcontractors. One Massachusetts builder took the extreme position that it is always a mistake ever to assume that any subcontractor will arrive on time. He, and many others, suggested firing anyone who does not arrive when scheduled. This is an obvious solution, perhaps, but one that is seldom adopted. You simply don't want to have to go to the trouble of hiring someone else, and the subcontractor assures you that he will arrive at a new time, without fail. Even when he still doesn't show up, and doesn't phone—construction people sometimes seem to have overlooked the invention of the telephone—hope springs eternal. And so, promises heap upon promises until in final exasperation you say, "The hell with him," and hire another subcontractor—who also may not show up when promised.

Another common piece of advice offered by owner-builders is to keep a close watch on the work of subcontractors; never assume that, because they are professionals, they know what you want or even possess the skills you are paying them for. Many of the so-called skilled workers in a construction team may have been working at their jobs for only a few weeks or months. Unless you hire union labor, there is no guarantee that those working on your house have achieved any level of competency in their trade. Short cuts, errors of judgment, hastily concealed poor workmanship, and downright blunders are more likely to occur when the owner-builder is not at the building site to oversee construction.

The only control a self-help builder has over the quality of a subcontractor's work is the purse string. A number of owner-builders confessed that they made the error of paying a subcontractor for work before it was finished to their satisfaction. Whether an entire job—framing, rough plumbing, roofing—is contracted at a flat fee, or whether a contractor is working on an hourly or weekly basis, the final payment should be withheld until the work has been thoroughly examined and approved. Where codes are involved, the work should be inspected before payment. This is one of the most often repeated pieces of advice from builders.

Subcontractors may attempt to tack additional costs to a contracted price that is not set down in writing. Many independent contractors work with a simple verbal agreement on the price for a job. Owner-builders are within their rights to refuse to pay anything above the agreed-upon price. In cases where the subcontractor purchases materials to take advantage of wholesale prices, owner-builders should make arrangements to have the bills forwarded to them rather than to the subcontractor. This way, the builder is sure that they are paid. It is not unheard of for self-help builders to have a mechanics lien slapped on their homes when a subcontractor disappears without paying his bills. Subcontractors such as electricians and plumbers, who supply the materials for their work, should specify the type and quality of the materials to be used. If this is not done, the subcontractor may choose the cheapest materials; the money saved is extra profit on the contract.

If these accounts of the practices of subcontractors sound unfairly disparaging, it is because all of the evidence I have gathered supports the conclusion that subcontractors can be one of the most nagging irritants to the owner-builder. It is no doubt true that most subcontractors are courteous, fair, and conscientious, and their work is performed satisfactorily, but the owner-builder does not get to meet enough of them. Because low cost is frequently the main consideration, and because the owner-builder is a small, one-shot piece of business, he or she gets more than a fair share of the incompetents and cheats. One way of avoiding this is to seek not the lowest-priced subcontractors but those with the best reputations among other owner-builders. This is where self-help builder networks can be extremely helpful; a recommendation from another owner-builder is perhaps the best guarantee you can have that the work of a subcontractor will be satisfactory. A highly recommended electrician or plumber may cost more, but as one self-help builder put it: "Get the best labor you can find if you want to be happy with the work they do. Good labor is never cheap; it's like the materials you buy—you get what you pay for."

Difficulties with subcontractors lead many owner-builders who originally planned on handling only a few systems themselves to take on more of them as construction continues. It can sometimes be more of a burden than one bargained for.

One builder reported that after waiting in vain for a backhoe operator to show up to excavate for his septic tank, he decided to hand-dig it himself. "It took me five days," he recalls, "and after I finished it I told my wife I wanted to be buried there because I had spent the best part of my life in that hole."

Another owner-builder, a neighbor of mine, perhaps illustrates even better how problems with subcontractors can be solved. He brought in a team of masons to lay the brick veneer on his house, but became increasingly dissatisfied with the quality of their work. I examined some of the mortar joints he complained were not uniform and found them to be a bit too thick here and too tight there, but I observed that no one would ever notice it. But he was more of a perfectionist than I, and he finally fired his subcontractors and proceeded to tear out the irregularly laid bricks and finish the masonry himself, with the help of his wife. They moved on to the fireplace and chimney, and completed them. Then, getting carried away with their new skills, they tiled the floors of several rooms with thin slabs of polished marble they had salvaged from a tombstone works. The only problem was that firing their subcontractors and doing it themselves cost them the better part of a summer. That is usually the give and take in self-help building: time for quality, time for money.

When confronted with the unavailability of professional help at the time it is needed, owner-builders often resort to the flexibility inherent in the self-build concept. Many masons, carpenters, roofers, sheet rock finishers, and painters are quite willing to moonlight on owner-built houses during the evening hours and weekends. They are not always able to bring their equipment with them, but in many cases it can be rented or borrowed. This arrangement usually gives the self-help builder an opportunity to master a skill more rapidly by working along with the professional, or at least by closely observing professional work.

When skilled help is not readily available at the right price, the owner-builder can sometimes resort to unskilled labor to push the work of a system along at an acceptable tempo. For example, almost every owner-builder I have interviewed admits that concrete work and laying concrete blocks are two of the most wearisome, physically exhausting jobs in the entire repertory of house systems. Much of the hard work in these operations consists of shoveling sand and gravel, pushing wheelbarrows, lifting blocks up on scaffolds—unskilled manual labor.

By breaking down the concrete and blocklaying systems into skilled and unskilled subsystems, an owner-builder may be willing to take on a job that otherwise seems overwhelmingly laborious. Skills must first be learned by practicing on a small part of the project, but once minimal proficiency is gained, the full job can be completed by using low-cost, unskilled labor. A man/woman owner-builder team, for example, can hire one or two high school or college students to do the mixing and carrying jobs while they finish the concrete work or lay blocks. Organizing a task this way makes it manageable and pleasurable. Such other systems as roofing are no more than incremental repetitions of simple tasks, and can also be performed by unskilled labor. Once the initial row of asphalt shingles is positioned and chalk lines placed for subsequent rows, finishing the roof consists simply of nailing down shingles—a job that takes little skill. During summer months students are nearly always available at the minimum hourly wage for such work. (But be certain you have liability insurance.) The time saved is almost always exponential: two persons working at construction complete a job three times quicker than one. This is one reason why a team of carpenters and their helpers can frame a house in three or four days, whereas this job took me, with only occasional help from my son, months to complete.

THE EXPENSE OF "FREE" LABOR

Labor that is free—offered by friends, relatives, and neighbors—can be a boon, but then again it is sometimes not so free. Managing and supervising hired, professional help is difficult enough for the amateur—they *are* the professionals and they know how to do it, even if it isn't the way you want it done. In a pinch, however, the owner-builder can always fire them and retain control of the building project. However, when relatives or friends are working for nothing and their work proves unsatisfactory, the builder may be faced with a ticklish diplomatic problem. Friends and relatives cannot be summarily dismissed; feelings are involved, egos and friendships at stake. When Uncle Wally, who has been a carpenter for thirty years, insists on doing it the way he has always done it instead of the way the blueprint dictates, what can you do? When the weekend costs of food and drink for the "free" help that shows up with family and kids exceeds the cost of hiring unskilled laborers, do you invite them back? And when you spend more time tearing out and redoing the shabby work of free help, is it really worth it? These kinds of questions lead a number of owner-builders to conclude that there is no such thing as *free* labor.

Decisions on whether to hire subcontractors, skilled or unskilled labor, or to solicit free help will ultimately depend, as does every other decision in building a house, upon the owner-builder's personal style. The demands placed on help to do exactly what is wanted, the standards that are set, the care taken in hiring help, the abilities to work comfortably with others, the schedule demands of a particular system, the skill in managing finances—all will figure in each decision. I have already suggested that the more autonomy owner-builders retain over their enterprise the more rewarding they will find it. When others work for you, there is an inevitable loss of some control over the building process. In many cases, you are quite willing to give up some of this authority in order to expedite work. In spite of the pleasures of completing a job yourself, there are some systems the builder simply wants to get finished as quickly as possible. I experienced a rush of pleasure when the septic tank contractor arrived at my house with a backhoe and proceeded to gouge deep trenches in the hard clay. I had never considered doing this work myself, and I was anxious to have it finished. A few days later, when I inspected the neatly graded field where a concrete basin and plastic lines lay buried to dispose of my future waste, I felt as proud as if I had done the work myself. Many house systems are like this: once the decision is made to subcontract them, the main concern is to see the work done as quickly as possible—but done satisfactorily. This is why the interminable delays experienced with subcontractors are so exasperating, and also why poor workmanship grates so much. "It I wanted this kind of lousy work," one owner-builder complained, "I didn't have to wait for it— I could have done it myself."

How well you are able to work with others will also determine whether you bring in outside helpers and how comfortable it is working with them. When an owner-builder hires a subcontractor, he or she becomes a bystander to the work; the subcontractor's team operates as an autonomous work system which performs according to its own methods, rules, and inner dynamics. As the team moves from job to job, its members adapt only slightly to the changed environment; the way they work together and the techniques they use are predetermined by the complex web of traditional work habits, personal idiosyncracies, interpersonal relationships, and informal rules of the trade. Sometimes self-help builders can hire themselves out as unskilled helpers to the subcontractor, a strategy that enables them to watch over the quality of the work, acquire skills first hand, and save some money. In such a case, the owner-builder fits temporarily into the work team, but as a supernumerary rather than as a full member of the cast.

When an owner-builder hires skilled or unskilled laborers to work as part of the team he or she devises, the interpersonal relationships are altogether different. Whether this team functions as a unit will depend on the owner-builder's ability to socialize work. Almost all vocational work is public and social; indeed, the ability to adapt to the social demands of a work environment dictates the success of the work experience. An avocation such as building a house, however, does not need to be social; an owner-builder can choose to work entirely alone, staying aloof from the work processes of those subcontractors who are hired. As in other aspects of life, there are many house builders who are loners—independent, self-sufficient, determined to impose their will alone on both house concept and materials. The self-help building movement attracts this type of personality, individuals who express themselves most comfortably in a private rather than a social context. It is compatible with the artistic personality, for most of the traditional arts are solitary expressions of human creativity. Writing, composing, painting, and sculpture, as opposed to such collaborative arts as theater and dance, are all practiced in isolation. Constructing a house can be a fulfilling experience either way, privately or socially. Those who are gregarious will find in shared labor, hired or donated, the same kind of camaraderie and companionship found in sports, where belonging to a team transcends all other personal considerations. Straining one's muscles in a common endeavor, the work equivalent of a tug-of-war in game playing, can produce deep feelings of shared work-intimacy. Perhaps the fusion of work and play is nowhere more evident than during those times of cooperative effort when the coordination of muscle and limbs produces a tuned, work harmony. It is felt even more intensely by terpitects than by those helping out, because it belongs to their personal endeavor in producing shelter.

Straining one's muscles in a shared experience, such as the concrete walk these Self-Help Enterprise members are building, produces feelings of work intimacy, one of the pleasures of owner-building. (Photo: George Ballis)

OWNER-BUILDER NETWORKS

Happen upon an owner-built house, and the chances are you will find another one close by. This simple truth represents one of the enduring strengths of self-help building and also its greatest hope for making an even greater impact than it has on the way Americans house themselves. Although the desire to build your own house is an individual impulse, the process of building—from inception of the idea to finished house—is intimately connected with other self-help builders. Without interconnecting networks of owner-builders, the number of self-built houses would be only a fraction of what it is.

It is not uncommon to drive up a suburban street and find that half or more of the houses are owner-built. Once self-help building gets in the air in a community it can be contagious; building fever can settle into a neighborhood for years and spawn ever-spreading colonies of owner-built houses. The lines of communication between builders in such colonies are entirely informal; a member of a family agrees to give advice or help to a cousin or relative who is interested in building; fellow workers discuss building a house over lunch or drinks; neighbors stop by to observe neighbors building. From such loose, disorganized, accidental exchanges, decisions to build are made, encouragement advanced, tools borrowed, advice solicited, labor offered, sources of materials found, problems talked out.

It is small wonder that businessmen and bureaucrats refuse to believe that owner-builders are anything more than masses of uninformed, unskilled individuals, like so many billiard balls rolling in spheroid isolation, because self-help builders are part of no visible organization. The kinds of systems that professionals recognize and understand are hierarchical ones with direct lines of authority and structured modes of exchange. The driving force behind commercial systems is self-interest, with its demand for the integrity of organizational controls to prevent profits from falling through the cracks. There is no place in such systems for qualitative, unpredictable, human links like neighborliness, friendship, altruism, gregariousness, or curiosity—the sorts of links that tie one owner-builder to another.

These are wasteful, inefficient, and time-consuming by business standards, but by human measurement they not only produce the warmth and pleasures of interpersonal contact, they also get houses built.

The willingness—the desire—of one self-help builder to aid another is probably rooted, in part at least, to the same kind of cohesive force that draws together any group sharing a passionate interest. The sudden discovery of music, dance, tennis, running, or even dieting can strike an individual with an almost religious fervor. Like someone who has undergone a conversion experience, there is a desire to discuss the topic, to explore with others its profundities and mysteries, and to share its bliss. In the case of the owner-builder, this desire to join with others in a creative infatuation is augmented by the need to maintain and sharpen newly acquired skills. Just as many self-help builders decide to construct a house of their own simply because they have the ability to do it, those who acquire house-building skills have a need for finger exercises, etudes in carpentry and masonry to keep in practice. One way of flexing those skills is to help friends or neighbors build their house.

Owner-builders are not limited to receiving information and aid from other self-help builders, however; their flexibility and autonomy also enable them to tap into the efficient and well-organized systems of the housing industry and its support structures. All around them, commercial buildings are being constructed, contractors are putting up houses, building suppliers are pushing construction materials. To bankers and businessmen, professional construction workers may be merely necessary cogs in an industrial machine, but as individuals they are the same kind of person as the owner-builder who wanders onto a construction site to chat with them. They are likely to be open, friendly, and willing to offer advice or to help solve a problem. And because construction practices are so standardized, owner-builders who look around a professionally constructed building site will see a model of the very structure they are building. They can observe how a house is framed, how wiring and plumbing are integrated into its structural form, how house systems are put together. The builder can become a student at the feet of professionals by merely observing and

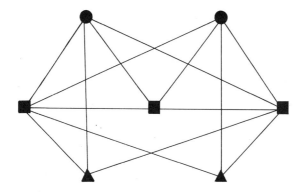

OWNER-BUILDER NETWORKS

■ Owner-Builders

● Building Materials Suppliers

▲ Family, Friends, Work Associates

This schematic diagram illustrates how building information is exchanged informally through owner-builder networks. In spite of the casual spontaneity of such networks, they represent one of the unique and enduring strengths of the owner-builder movement.

noting. An owner-builder walking through a half-finished house *sees* construction as few others do. He or she closely notes construction details that even the professional ignores because they have become so habitual. A building site can be as important a source of information and as valuable a network as are other owner-builders.

One other network serves the owner-builder—labor and materials suppliers. The professional construction workers who are hired are a source of information and advice on who is the most reliable supplier, what kind of supplies to buy, and where to find used or salvaged materials. And suppliers are often the best source of information on where to find laborers. Many of the salespersons at lumber yards, hardware stores, and building supply centers are former construction workers who not only know how materials are used in construction, but are motivated to ensure that they are properly installed—a satisfied customer will return. Large discount building supply centers are designed

HOW TO ORDER 2-SECTION EXTENSION LADDERS

For each extension ladder size there is a range of heights that can be reached safely. Using too long a ladder (even when completely closed) to reach heights lower than those recommended within the range given for each ladder size can be dangerous. Check chart below to determine what extra ladders may be needed to reach desired heights.*
NOTE: If you are going to climb off of ladder onto roof order next size ladder in order to have something to hold onto.

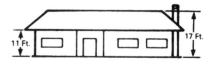

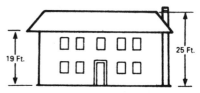

(Dimensions shown above and below are for typical house construction)

Reachable height range	Ladder size	Extra ladder needed to reach lower heights		
		Extension	4-in-1	Stepladder
4 to 11 ft.	14 ft.	....	....	....
5 to 13 ft.	16 ft.	....	....	....
7 to 17 ft.	20 ft.	....	....	....
9 to 21 ft.	24 ft.	....	....	5-foot
11 to 25 ft.	28 ft.	....	....	6-foot
13 to 29 ft.	32 ft.	16-foot	5-foot	8-foot
15 to 33 ft.	36 ft.	16-foot	6-7-foot	10-foot *
17 to 36 ft.	40 ft.	20-foot	8-foot	....

Mail order houses offer a wealth of free information. The Sears and Ward catalogs have been a source of supplies and building information for thousands of owner-builders. (Reprinted courtesy of Sears, Roebuck and Co.)

for do-it-yourself merchandising and they are eager to offer as much installation advice as possible. Manufacturers of construction supplies actively promote amateur building by publishing do-it-yourself pamphlets, brochures, and booklets, and by encouraging salespeople to offer construction advice.

The self-help builder will soon come to realize that there is an eager ally in the building supply industry. The reason is that owner-builders are now a significant segment of a huge consumer market. For every house that is owner-built, another is being renovated, restored, or rehabilitated, and for each of these there are dozens that are in the process of being spruced up, repaired, or maintained, all by their owners. This adds up to an annual expenditure in billions of dollars by Americans on home construction, repair, and maintenance products; manufacturers and retailers compete actively for this market. If the home builder needs information on how to install one of their products, they will do their best to be accommodating.

NINE

RELATIONS:
THE OWNER-BUILDER'S
FAMILY

Most Americans
who choose to build their own houses are married. They are
not unlike home buyers in this respect; although more single
persons are buying houses every year, 80 percent of the nation's
new housing is sold to married couples. This means that home
building is not an individual act for most self-help builders, but
a cooperative family endeavor; therefore, all decisions on planning and executing the project must be made within the context
of the family constellation. Building a house takes most families
at least a year, and often much longer. During that time, family
routines are severely disrupted, social obligations are abandoned or curtailed, and there is virtually no time for any activity
but construction. This intensity of commitment to a commonly
shared goal is a great strength, and it consolidates family ties
as no other joint activity can. But it must be truly shared if it
is to be a creative experience for the family. If not, children,
for example, can become insecure and resent the construction

165

project for coming between them and their parents or for breaking habitual routines. A Maryland owner-builder, who built a split-level contemporary house with the active support of his family, summed up the critical importance of solidarity: "The pleasure of self-help building depends on the attitudes of the entire family. If it is looked on as a hardship that father punishes the family with, then it is hell. If it is a joint project full of adventure and combined fun, then it is a joy." The best way for the house to become a joint project rather than father's—or mother's—scourge, is for the entire family to participate in all decisions. This can begin at the planning stage when space allotments are made. A family council is perhaps the best way to organize such planning sessions. Each member of the family can voice his or her opinions, counter objections, and work out compromises in an open discussion, where everyone has the right to speak out. Even small children can have a voice in planning those parts of the building project that will affect them—their own bedroom, for example, or family and play areas. However, to make such sessions work, parents must be willing to listen to children and heed their desires.

The actual construction period can be a time of both fun and anxiety for children. Kids of all ages are able to take part in house construction and should be encouraged to do so. Teenage boys invariably finish the house with a firm resolution to one day build their own houses, but it is also satisfying for girls of all ages. There is no better way of breaking down female role stereotypes than by having young girls don work clothes and hard hat and pitch in with hammer and paint brush. The construction period is hard on children, however; it disrupts the regularity and order of family life, often at a time when children are most in need of security and stability. Father and mother do not spend with them the time they are accustomed to, and when parents do find a few moments for the family they are often weary and in no mood for supplying support or entertainment. Parents can also bring the frustrations and annoyances of the building site home with them and inflict them on the children.

If the family moves into an incomplete house, conflicts can be severe. Teen-agers who are dating can be particularly un-

Teenage members of owner-builder families invariably enjoy the fun and excitement of building. And there is no better way to break down sex role stereotypes than by having teenaged girls participate actively in building or rehabilitating a house. (Photo: National Archives)

reasonable (to their parents) about appearance, status, and clutter. They may look upon living in an unfinished house as socially declassé, and prod their parents with the kind of remark I heard one teen-ager make: "When are we ever going to get this pigpen finished?" When a building project stretches on, it is not always easy to maintain a spirit of excitement and adventure, but if the self-built house is to be a success, the effort must be made. One family kept morale alive by giving a special party each time a new section of the house was completed. When the bedroom of the six-year-old daughter was finished, kids in the neighborhood were invited to a "sleep party." The teen-age daughter celebrated the completion of the family room with a disco party, and the parents threw a series of indoor and outdoor beer busts during various stages of building. Not only did the entertainments encourage the entire family to associate house building with pleasure, but they also gave neighbors and friends an interest in the construction project and an understanding of its difficulties and rewards.

With young children, parents might use the Tom Sawyer strategy of turning work into play. Recall that Tom induced his friends to whitewash a wooden fence for him by convincing them that it was important, responsible, artistic work that only the gifted could be entrusted with. By tailoring the task to the child, by allowing kids to do real, grown-up, creative work—not just the clean-up jobs—parents can turn construction into work-play for their children.

Strategies such as these are important in maintaining children's involvement in the building process. It is difficult for them to take the long view. They want instant gratification; they want the house finished now. A year or two in the future is forever. When construction tasks or lack of money interferes

One way parents can keep children happy while home building is in progress is to maintain family rituals. These children have their room and playthings as close to normal as possible while living in a house under construction.

with what they perceive to be essential needs they can be emotionally crushed or resentful. The house can become a wicked stepmother who deprives them of all of their fun. A recognition that what may seem like trivialities to adults can be earth-shattering for children, and a willingness to discuss complaints and resentments openly and candidly can help resolve these kinds of conflicts.

It is easy for parents, whose attention and creative thinking are concentrated on their building project, to overlook the domestic activities that give cohesion and stability to a family. Going out to a restaurant, watching favorite television programs, taking kids to the movies or sports events, going shopping together, visiting friends or relatives—these family rituals are even more important during the building effort than ordinarily. They must remain a part of daily life even if they involve a degree of inconvenience; they tie the present, impermanent, disruptive, disorderly period to more familiar and more secure times.

MARRIAGE CONFLICTS DURING CONSTRUCTION

The impact of home construction on a marriage can best be summarized by a comment repeated by numerous owner-builders: "If a marriage can survive building a house, it can survive anything." Once a husband and wife decide to build, they make a major commitment of economic, physical, and emotional resources. Although this decision is seldom made lightly, the couple rarely realizes the magnitude of the job they have undertaken. "We had no idea what we were getting into," is one of the most commonly heard remarks made by house-building couples.

Traditional marriage roles split domestic work into sex-defined categories: the man stays out of the kitchen—that's women's territory—and the wife keeps out of the shop. Taking care of the kids is women's work; taking care of the finances is a man's job. These traditional roles no doubt still apply in many marriages, but increasingly they have broken down completely, and domestic tasks are shared with a degree of equality that was unheard of a generation ago. Men cook, clean house, and take care of children; women hold jobs of their own, and plan and manage household finances.

It is less common, however, for even the most independent of contemporary working women to have had a great deal of interest or experience in the kinds of construction skills demanded by house building. When a couple makes a decision to build, therefore, the woman may be much less prepared for the physical work involved. Women do adapt and learn, however, and become as involved in the construction process as men. In the planning and management systems of a house, it is often the woman who assumes authority and makes the major decisions.

Building a house together does not present to a couple any problems in interpersonal relations that they have not already dealt with in their relationship, but it does intensify them. The number of joint decisions that must be made is probably unprecedented for them. Each of these decisions, whether it is a big one, such as how many bedrooms the house will have, or a trivial one, such as the angle at which a nail ought to be driven, is a source of potential conflict. The ability to compromise is essential if constant quarreling is to be avoided. If compromises cannot be struck, one member of the team ends up reluctantly deferring to the other, and a smoldering anger can result. The clockwork of interpersonal relationships—the needs, defenses, and adaptations of each individual—is wound up tightly during the building period, and there are numerous occasions for the explosive release of emotional energy. Avoiding these, and channeling stress and anxiety into creative outlets is one of the main challenges house building presents to an owner-builder couple.

If there are unresolved conflicts in a relationship, it is possible that disagreements about the building project may in reality be disguised attempts at establishing a position in an ongoing dispute that has nothing to do with the house project. It is well for couples to recognize that building a house together has the potential for new intimacies born of shared work and creativity, but also that it can open and fester old wounds. Perhaps the greatest number of unresolved conflicts involve struggles for power and dominance. Who wears the pants in the family? Who is responsible for raising the children? Who handles the money? Who is superior and who is inferior? Who is forceful and who is passive? Who is assertive and who is submissive? Who is confident of his/her identity and who is not? If couples are aware at the outset that queries such as these will be shouted, unvoiced, to the rafters innumerable times during the construction process, they will be more likely to hear and respond to these vital questions over the din of the sounds of building.

Several man/woman teams from my sample of owner-builders offered advice on how to complete a house in an even happier state than you were at the outset. A couple building a house in California talked of the importance of work parity between a man and a woman. She commented:

> Equality, that's the secret. You share work equally and you share responsibility equally—and that means sharing the blame for mistakes and not laying a guilt trip on your partner. You don't have to match an hour's work for an hour's work, but you do have to recognize that while the guy's framing a wall the woman is doing equal work by keeping the baby and dog out from under his feet. Or her going off to the hardware store to buy nails is as important as his hammering them into the roof.

A Detroit fireman stressed communication:

> You've got to be willing to talk things out. Whenever my wife and I disagreed on something—and believe me that happened almost every day because my wife's a strong-minded woman—we'd take a break and sit down and talk it over. At first, these were more like fights than talks, but after a while we decided, hey, this doesn't make sense, yelling at each other over some little thing like whether an electrical outlet goes a foot to the right or left, so we made it a rule that if I gave in on this one, she'd give in to me on the next, and that worked out just fine, and our talk breaks got to be really talk instead of fights.

A mechanical engineer, who built a contemporary saltbox house in Connecticut, spoke of the importance of maintaining some semblance of a normal married life during construction:

> It's easy to become so wrapped up in the day-to-day grind of building that you forget that you're still living your lives together and raising a family and the kids still need braces and the dog needs shots. In order to keep our sanity we just periodically called a halt to building and went out for a night on the town or got drunk and made love or took the kids out for a picnic—we pulled life back to some kind of perspective. Then when we went back to work again the day or two we missed were more than balanced by the renewed energy and sense of excitement we brought with us to the job.

A Virginia mother balanced her one-year-old daughter on her knees as she talked of husband/wife relationships during construction of her house. Nearby, her husband shoveled gravel for a batch of concrete that would become a walk in front of their colonial ranch house.

> We had a lot of trouble in the beginning. He was the expert and I was just a housewife with a new baby, and I felt that the house wasn't mine at all. He decided on building it pretty much on his own, and his parents put up the money.
>
> Right after it got framed, I fell into a depression. Larry was working a full day on his regular construction job, and then working all kinds of hours at night and all weekends on the house, and I was sitting in the apartment with the baby. I got so I hated the house and was sorry we ever started it. After a big blow-up about it one day, we visited a therapist. She was really great; she told my husband that if we were ever going to be happy in the house it had to be *our* house—not just his. And she suggested that instead of sitting home, I go out there, take the baby with me, and work along with him.
>
> So I did just that. I prepared food for us and we'd take it to the building site, and I really began to enjoy the work. I did everything! You wouldn't believe that somebody who had never even held a hammer before in her life could drive a ten-penny nail into a 2 × 4 with three whacks.
>
> Working on the house gave me more self-confidence than I had ever had before, and it did a lot for our marriage. When

Larry saw me doing "man's work"—and liking it—a lot of his attitudes about me changed. I don't think he ever really thought of motherhood as being worth very much—not like going out and working at a job, for instance—but the work I did on the house was important and I was a partner in that. And by the time we finished, I was a real partner; I was picking out more of the paint and appliances and doing more of the decorating than he was. I honestly think that if I hadn't gone to work on the house, our marriage wouldn't have lasted.

The last word came from a sixty-eight-year-old former postal worker who built a retirement home in Florida:

> Me and my wife settled problems in building the house the same way we settled all problems during thirty-nine years of marriage—when she complained, I gave in. Never had no trouble at all.

WOMEN TERPITECTS

One of the most frequently asked questions directed to owner-builder couples is, "How much of the work did the woman do?" The unstated assumption behind this query is that women, naturally, are not able to compete with men in doing construction work. They are not strong enough; the work is too physical; they don't have the mechanical aptitudes for it; construction work doesn't interest them. My research into owner-builder behavior has taught me that women are not only able to do all tasks involved in house building, but they do them with a frequency that would disarm the most macho construction worker. Thousands of women work alongside men as equals, learning and mastering each job as it comes along. Women take pleasure in every kind of craft, including such traditional male skills as carpentry, masonry, plumbing, and electrical work. It is not unheard of for a woman to build most of a house herself while her husband or companion works overtime to finance it. One Tennessee woman, who did everything but the framing on her house while her husband held a full-time job, commented, "One of the greatest rewards of the entire project was

Women are increasingly building, restoring, and rehabilitating houses. The owner-builder movement has demonstrated that women are able to master all the skills of home building— and to enjoy doing them. Here a Self-Help Enterprise builder proves that electricity holds no unsolvable mysteries for the informed amateur. (Photo: George Ballis)

when the electrical inspector complimented me on the wiring I did."

These are not extraordinary individuals; they are women who have broken through the role stereotypes that define acceptable female work behavior. One woman self-help builder stated her attitude toward socially-enforced roles this way:

> OK; if they say they aren't hiring women carpenters on construction jobs, that's one thing. That's their business. But this is my house, and I'm the one who decides whether or not I'm able to do the work and what kind of work I want to do. The decision is up to me and my man. I don't care what kind of work is traditionally "women's work" or "men's work"; what I'm interested in is whether it's *my* work.

Those women who do not choose to do the more physical parts of house construction are invariably closely involved in planning, management, record-keeping, and purchasing. Some women prefer light construction jobs: painting, staining, wall-papering, paneling, carpet and tile installation, and decorating. A North Carolina woman explained how her attitudes about construction work changed during the house project.

> At first, I was trying to prove myself—and impress my husband—by doing everything, even jobs that left me exhausted, like concrete work. But then I decided that I really didn't have to prove anything to anybody; I could just do the jobs that I really liked doing. As it turned out, these were the more artistic things like painting, staining, decorating, making draperies and valences, things like that.

It is much less common for single women to build their own homes, but this may be changing. One of the dramatic events that has occurred in domestic housing is the sudden appearance of unmarried home buyers in a market traditionally limited to married couples. The National Association of Home Builders reports that single people have been buying as many as 20 percent of the homes in some subdivisions. In the case of townhouses and condominiums in some areas of the West coast, 80 percent of the buyers are single and 50 percent are un-married women. The reasons are both economical and psycho-logical. The economics of a townhouse makes a lot of sense to a single woman; it allows her to enter into the constantly-in-flating housing market for the least amount of money, since townhouses and condominiums often sell for half the price of a single-family house. It also gives her access to a community of single people like herself.

With so many women discovering the economic and personal advantages of owning a home, it is only a matter of time before more of them will learn the pleasures of building a home. Single women are presently at a disadvantage in self-help building because they do not have the ready access that men have to construction information channels in our society. Men can easily exchange building advice and information at work and among friends and neighbors who have either built houses or who are construction workers. Self-help builders are mostly men and couple teams, and the owner-builder networks which function

as major sources of encouragement, skill-acquisition, information about materials, and help of every kind, may not be readily
available to a single woman. Entrance into these networks,
which occurs spontaneously with men or man/woman teams,
is likely to require a degree of effort for the unmarried woman
whose work and life style may be remote from construction of
any kind.

Nevertheless, I think it is inevitable that women will in ever-
increasing numbers take into their own hands the responsibility
for creating their own shelter. The feminist movement, together
with legislation banning job discrimination by sex, has opened
new opportunities for women in traditional male vocations. As
the barriers come down and women enter increasingly into the
previously all-male world of construction, as women lose feelings of inferiority about doing "male" work, they will seek out
those avocations such as house building that have previously
been denied them.

It is something of an irony that the home, which epitomizes
the feminine principle in all societies, continues to be an artifact
of male creativity. The feminist movement has provided an
impetus for women to join together to learn how to build a
house. In California, where many social revolutions in this country are born, women are making an entry into the previously
all-male business of general contracting, by first building their
own homes, often with the help of other women. They have
also formed owner-builder cells which function as teaching and
information resources. When a woman has learned a few construction skills, often by taking classes at a local tech school and
by working on an owner-built home, she is then in position to
seriously think about constructing her own house. Many more
women in cities are already involved in rehabilitating and restoring old and rundown houses, lofts, and apartments. But
these are only a faint hint of the potential for women's self-help
building. If the tenacity and singleness of purpose which has
driven women, often against seemingly insurmountable odds,
to open day-care facilities, abortion clinics, and counseling centers, could be directed toward women's self-help building there
is no predicting the number of houses that could be built by
women alone.

One self-help builder, a general contractor who uses women crews and who teaches house building on the West coast, discussed the problems and rewards of house building for women.

> How to put things together, how to build a house, are not such big problems. But as women, we are not exposed to the way of thinking it takes to do these things. In school, I didn't get to take shop or auto mechanics—the practical things to create the thought pattern to do something like build a house. If I had been a boy I might have become involved. Now, I've had to work double time to learn these things. What I try to do for women who work with me is demystify the process of construction. It's amazing how someone's life can be changed by just building a wall.

Single women who build their own houses are really doing work that is no different from that done by the thousands of married women who construct homes alongside their husbands. The difference—and it is a significant one—is that the single woman must initiate all action on her own; there is no man to seek advice from, to place one's trust in, or defer to. This is an entirely new attitude for many women, whose work experience has always been dominated by male authority, and it can be challenging. Women who still hold to more traditional roles and who are reluctant to plunge totally into what was until recently the all-male world of construction might do what a number of men owner-builders do: hire a retired contractor or builder or master carpenter who is between jobs to supervise the construction of the house shell, and then finish the interior yourself. With the house dry and closed in, the woman self-help builder is in a position to test her own skills in a leisurely fashion and to discover for herself, at her own pace, the pleasures of terpitecture.

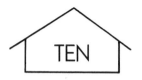

TEN

GOVERNMENTALITIES:
CODES, HOMESTEADING,
AND REHABBING

The two points at which the government comes in contact with the owner-builder most directly are in collecting taxes and in enforcing building and occupancy codes. Of the two, codes are often the more oppressive. I did not have to deal with building code inspectors when I constructed my house; therefore, my objections to them are not as strident as those of many owner-builders who had to put up with the pettiness and frustrations of a bureaucracy that too often serves the owner-builder poorly. Codes are a part of the owner-builder equation in many communities, however, and are best met the same way other subsystems are approached—by simply accepting them as a given. It would be much easier if they did not exist; but they do, and they must be lived with. To allow them to chafe the spirit is to turn the pleasures of building into anxiety or anger. Most of the owner-builders I queried had little or no difficulty with building codes. They built according to code requirements, arranged for the necessary inspections, found in many cases that inspectors were human and helpful, and then got on with their building.

I think that it is important for the owner-builder to adhere to the basics of building codes, even if he or she lives in an area where codes are nonexistent or are not enforced. I followed all codes, simply because they gave me the security of knowing that I had a minimal standard for safe building. In most cases, my construction was far in excess of code requirements, and I think this is true of most self-built homes. Owner-builders are interested in quality work and codes are one of the assurances of minimal quality.

In cases where codes appear to be irrational, where they exist merely to serve the self-interest of professional tradesmen, one must be philosophical about them. They are what Thoreau called "the friction in the machine," the injustices that are the inevitable part of every institution. To allow them to disturb us unduly makes them like the mote in the eye that reduces all beauty to a nagging, tearful blur.

Building codes have produced the most friction in places where there is a conflict of life style rather than building style. On the West coast, self-help building has become for many not simply an alternate way of producing shelter, but the keystone of an alternate life style. Many California, Oregon, and Washington builders are the squatters of an affluent society, not people who are so poor that they simply take over a piece of public land and start building on it, but idealists who have adopted a philosophy of voluntary simplicity. They have rejected the materialistic values of our consumer society in favor of the self-sufficiency of Thoreau. Their life style is based on limited needs and reduced consumption. If needs can be trimmed enough, you can be released from the tyranny of meaningless work performed only to buy the artifacts of a modern industrial society—house, car, television, novelty goods. If you can raise your own food, build your own shelter, and reduce your needs for the products of the marketplace to a bare minimum, you can do the kind of work that you find personally enriching and creative. The foundation of this program is building your own house. Because those who choose voluntary simplicity usually have little capital, and because this kind of life style prevents the accumulation of capital, building must be done on a shoestring. Construction is usually on out-of-the-way, cheap land which is marginal for any other use, and building materials are often recycled or salvaged. In order to avoid a mortgage, building is pay-as-you-go and add-as-you-go. House plans are often innovative, imaginative, or whimsical, designed as works of individual creativity. The results, when they are successful, are delightful expressions of the owner-builder's lust for self-expression. Often, however, they are not constructed according to the Uniform Building Code. For a sampling of these houses see the photographs in *Woodstock Handmade Houses*.

Building to code specifications is not only a case of making the house structurally sound; usually a noncode house, like most owner-built shelter, is overbuilt. The code demands the amenities that most of us take for granted in a house—indoor plumbing, electricity, heating, kitchen, bathroom, bedrooms. Alternate life style builders, however, are often quite willing to do without many of these comforts, at least temporarily, in favor of more primitive but perfectly functional systems. A privy or composting toilet may be perfectly hygienic and aesthetically satisfying for the rural area the house is built in; electricity may be rejected as wasteful of energy, or postponed until later. A wood stove can be more than adequate for heating and is eco-logically sound.

Yet all of these alternatives are rejected by codes. Building to code is not only expensive; in many cases it forces the self-help builder to construct systems that are not wanted for eco-nomic or philosophical reasons. It is the classic case of the state impinging upon the individual's freedom to live the way he or she wants to—or so those who oppose codes argue. Perhaps it is only in an affluent society such as ours that people could go to court to argue for doing away with toilets and central heat. The involuntary poor, as opposed to the voluntary im-poverished, want houses that are up to code; they want the same kind of safety, comfort, and convenience that the middle class has taken for granted. Blacks, Chicanos, and ethnic mi-norities involved in self-help rehabilitation housing do not go to court to demand less plumbing or less heat; their opposition to building codes derives from the demands by codes that work be done only by professionals or that such modern, labor-saving materials as plastic plumbing cannot be used. The notion that less is more is a philosophy that is attractive after you've had more. The dispossessed of this country—and the world—are still struggling for the levels of comfort, sanitation, and safety that standard building codes were designed to ensure.

It is, nevertheless, necessary that codes recognize the unique circumstances of the owner-builder. Ken Kern, Ted Kogan, and Rob Thallon argue the case against codes for the owner-builder in *The Owner-Builder and the Code*. They assert that codes fail to take account of the fact that self-built houses "traditionally have meager beginnings and contain long histories of additions, improvements, and remodelings," and that self-help building is often reparative rather than constructive. Yet owner-builders must submit complete plans to building departments before they start construction. "There is no allowance in the code," they write, "for the fact that the process of building one's own house is a piece-meal process, impulsive and passionate and often spread over a number of years."* Codes were originally designed to protect the users of houses from unscrupulous builders; they were not designed to protect builders from themselves.

Owner-builders have put forth a number of alternatives to standard building codes. United Stand, a group of northern California self-help builders, led the fight for the relaxation of codes—and failed. The story of United Stand is disheartening, for it demonstrates how impotent the owner-builder is when confronted with a well-entrenched bureaucracy supported by the power structures of the building industry. If owner-builders are ever to get relief from the more irrational demands of building codes it will have to come from cooperative action that carries the kind of political clout outlined in the next chapter.

*This book is in the best tradition of civil libertarian writing; its social conscience shows on every page. One sometimes feels, in this book and in Kern's other writings, however, that there is an inverse elitism, a moral superiority to the great mass of middle-class owner-builders whose desire to shelter themselves is really not too different from that of homesteaders, alternate life stylers, and rat-race dropouts who are Kern's constituency. Kern's impatience with middle-class values and middle-class owner-building fails to recognize that owner-building is a populist movement. One must be willing to accept the essentially conservative values of Middle America if any attempt is made to politicize self-help building.

When owner-builders as a group can muster the same lobbying strength of, say, the National Rifle Association, then it will be possible to bring about much-needed modifications to building codes.

GOVERNMENT AID TO SELF-HELP BUILDERS

Aside from enforcing building codes and collecting taxes, the role that federal, state, and local governments ought to play in self-help housing is a matter of some controversy. There are those who argue that self-help building is a viable alternative to public housing programs as a way of providing decent shelter to the poor of the world. They point out that owner-builder programs, initiated and financed by government agencies, have been successfully implemented in such underdeveloped nations as India, Peru, Colombia, El Salvador, Senegal, and Ethiopia. However, many of the practices which have been successful in other parts of the world simply do not apply to the United States. For the most part, these programs have supplied a minimal house that is not acceptable to even a poverty-level American family and certainly does not meet building code standards. In order for a government self-help program to be successful in this country, it must be capable of producing a house that is on a par with comparable housing in the community in which it exists.

The government self-help programs which have been attempted have met with only mixed success, largely because of difficulties inherent in any housing program administered by a bureaucracy. Most problems arise from the shift of autonomy from the house builder to the bureaucracy. There is ample evidence to demonstrate that the reason the owner-builder movement has steadily produced up to 20 percent of all single-family houses in this country for the past decade is that those who are responsible for completing the house, the builder and family, control the management and construction processes. This enables them to tap into informal networks for the supply of money, labor, materials, and information. Once a government

THE BALTIMORE STORY

How one city reclaimed its abandoned homes through homesteading and rehabbing. It is useless to attempt to rehabilitate a single house in a rundown abandoned neighborhood. Baltimore, Maryland, attacked the problem by rehabbing whole city blocks. Wire fences were thrown around row houses to prevent further vandalism and then homesteaders and rehabbers took on one house at a time. Sterling Street is one of Baltimore's showplace examples of what can be done when a city government and owner-builders combine their efforts. (Photos: HUD)

agency enters this equation, builder autonomy is compromised; all systems must flow through the bureaucracy. The economic advantage of owner-building is completely lost because the cost of maintaining the bureaucracy consumes the economies realized by the builder's sweat equity, if not to the individual builder, then to the nation as a whole, for the taxpayer ultimately pays for all such government aid. Red tape, delays, and the loss of individual choice further erode the creative pleasure of self-help building. The builder is confronted with experts who know how it should be done and who insist that it be done that way. John F. C. Turner has this to say about the difference between creative autonomy and agency control:

> To put it simply, building carried out by a large and hierarchically organized agency, whether public or commercial, provides little room for dialogue between people. All decisions are vertical and all operations are carried out by more-or-less unchallengeable order. When people are building for themselves, on the other hand . . . there is plenty of room for genuine relationships between the people brought together by the activity and, therefore, for creativity, pride, and satisfaction from the work itself. *Freedom to Build*, p. 145.*

Government self-help building programs have succeeded in this country when bureaucratic control has been minimal and individual initiative given free play. The most direct aid has been in the form of financing, mainly through the Farmer's Home Administration (FmHA), which provides low-income mortgages to rural families. Although it was not designed as a self-help program, it has made possible the construction of owner-built homes in rural areas. One reason why FmHA has succeeded in getting money to self-help builders, admittedly on a small scale, is that the relationship between bureaucracy and builder is on a one-to-one basis—the Department of Agriculture county agents who approve the loans, and in some cases offer technical advice, deal directly with the builder.

Although there have been numerous proposals to incorporate mutual aid owner-builder programs—groups of builders work-

*I am indebted to John F. C. Turner and Robert Fichter's *Freedom to Build* for many of the insights on self-help building offered in this chapter.

ing together—into government-sponsored low-income housing projects, such attempts have either remained small or failed, in part because of the apathy of the Department of Housing and Urban Development toward anything that is nonprofit. The one current exception is Self-Help Enterprises, a private, non-profit corporation which manages a mutual self-help program in the San Joaquin Valley. SHE was founded in 1965 by members of the American Friends Service Committee who were appalled by the deplorable housing conditions of Mexican-American migrant workers and who wanted to do something about it. The formula that evolved after several years of experimentation places eight to twelve families into a construction team whose goal is to produce a house for each family. Each team is organized into a cohesive construction unit through an educational program of several months which includes training in building practices, skills, and the use of tools. They also explore financing, credit management, taxes, insurance, and the responsibilities of owning a home. The training program is organized and managed by Self-Help Enterprises staff mem-

Of those who finish homes under the highly successful Self-Help Enterprise mutual home-building program, 60 percent are women. Here, women in the program join with men in constructing roof trusses. (Photo: George Ballis)

bers, but each team establishes its own set of operating rules and makes all decisions democratically by vote. In effect, the team becomes a tightly-knit network of builders and SHE technical advisors.

The actual construction phase usually takes eight to nine months and involves approximately 1,500 hours of labor by each family. Significantly, women complete 60 percent of the work on each house. The program has been underwritten by a series of private foundations and government agencies in the past, but in recent years Farmer's Home Administration has supplied funds for loans and staff support. There have been failures— some groups develop conflicts and disband—but SHE, in recruiting and screening candidates for the program, instills in them the importance of those qualities that support all owner-builder enterprises—dedication, hard work, patience, and pride. Since its inception, Self-Help Enterprises has produced 2,100 team-built houses in the San Joaquin Valley. Not only did the self-help builders produce homes for themselves and their families, many of them acquired new skills that enabled them to make the leap from field work to construction work. In recent years, SHE has also started a self-help rehabilitation program that has been responsible for upgrading from 150 to 250 rundown houses annually at a cost of only one-third of what a contractor would charge.

HOMESTEADING

One of the most publicized government programs involving the self-built concept is urban rehabilitation of abandoned slum houses, nostalgically dubbed Urban Homesteading. The idea was proposed in the early 1970s as a way to restore blighted inner-city ghetto houses and apartments and create viable neighborhoods. Urban homesteaders, like the settlers who trekked west in 1862 to acquire 160 acres of free land, would purchase an abandoned house from the government (a house that had usually been acquired when its owners simply walked away from it and refused to pay upkeep or taxes) and rebuild it. For as little as $1, the house would be turned over to the

A homesteader repairing brickwork in a Baltimore house. Many cities, unlike Baltimore, prevent homesteaders from working on their own homes by insisting that all labor be contracted to professionals. (Photo: HUD)

homesteader with the stipulation that it would be brought up to code standards within a given period, usually less than two years. When this was done, the house was deeded to the home-steaders. Loans would be provided to finance construction, but owners were to supply much of the labor themselves. The idea was originally conceived as a low-income program to provide home ownership for the kinds of families who had abandoned the rundown neighborhoods in the first place. There were some noteworthy cases where this was done, but the early attempts to implement the program failed for two reasons.

First, it was quickly discovered that rehabilitating a single house in a block of rundown, abandoned dwellings was fruitless; in a few years it would be like the rest of the houses on the block. It was necessary to rehabilitate neighborhoods, and for this to be done the city had to be willing to invest energy, creativity, and funds in a concerted attack on neighborhood decline. Once a neighborhood was turned around—when streets were made clean and safe and when residents took pride in their homes and their communities—then the boarded up, abandoned houses could be reclaimed and rebuilt.

Second, the amount of sweat equity that poor families would be willing or able to provide was overestimated. The initiative, determination, and self-discipline needed to master skills and manage scarce resources; the willingness to take on two jobs simultaneously, working and building; the patience to see a difficult project through to its completion—all the personal strengths every successful owner-builder must have—were frequently not available to would-be homesteaders. Some of the low-income families in rundown, inner-city neighborhoods were simply incapable of rehabilitating a house. Those who conceived of homesteading had no real idea of the amount of time and emotional commitment necessary for an amateur to strip a dilapidated house down to a shell and rebuild it. And without the self-building component in the rehab equation, the project was often not economically feasible.

However, after some initial shakeouts and false starts, after hustlers and speculators were identified and driven out, urban homesteading has become a reality in many of the nation's cities. But, like owner-built housing, it has become a middle-class rather than a low-income phenomenon. Nineteenth-century brick row houses and tenements in New York, Washington, Baltimore, Wilmington, Boston, St. Louis, and numerous other cities have been gutted and rebuilt into handsome, modern apartments and single-family dwellings. Rehabbing in the inner city is not very different from self-building in the suburbs. There are some pluses, but there are also a number of obstacles the owner-builder working from foundation up does not have to contend with.

Homesteaders have a house to begin with, but before they can start building, they may first have to tear out tons of inner skin, strengthen or redesign interior supporting walls and roof, shore up foundations, ensure the structural integrity of the house shell. They are then in approximately the same position as the suburban builder who has framed and placed siding on a house. All other systems—electricity, plumbing, roofing, drywall, carpentry, painting, and heating—must then be installed. The urban builder does not have the freedom from codes and enforced inspections that the suburban or rural builder may have. Most homesteading contracts demand that licensed contractors install electricity, plumbing, and roofing, and time limits are placed on completion of the rehabilitation work. Rehabbers who proceed on their own by purchasing an abandoned house and arranging their own financing and construction can, of course, avoid some of the code demands. Many rehabilitation projects are done surreptitiously, without a building permit, behind closed doors, room by room, over extended periods of time, financed paycheck by paycheck. Such bootleg rehabbers not only bypass codes and inspections, but also the assessor's inevitable tax increase.

The amount of rehabilitation that is under way in the rundown neighborhoods of the nation's large cities has increased dramatically in recent years, although not enough to match the flow of populations from the city to the suburbs. Census Bureau figures show that more families moved out of the cities in the 1970s than moved in. This is particularly true of blacks; because of fair housing laws and rising incomes, the number of blacks in the suburbs has increased 34 percent from 1970 to 1977. If this is the case, then who is homesteading and rehabbing the inner city?

For the most part, homesteaders are not suburban families returning to the city, but urban dwellers choosing to remain in the city. And these are not families with young children, but childless professional couples, unmarried couples, and "empty nesters" whose children have grown up and left. A study by the New York Citizens Housing and Planning Council concentrated on rehabilitation in parts of Brooklyn, the Upper West Side, and Greenwich Village and showed that the average age of rehabbers was thirty-one, compared with a citywide average age of forty-four. They also found that the average income was substantially higher than the city average, and that 70 percent had college degrees.

National demographic changes also play a part in the movement to rehabilitate older inner-city neighborhoods. The children of the World War II baby boom are now buying homes, but many of them have been priced out of the new housing market by spiraling costs. A homesteader's or rehabber's inner-city house is the only kind that many of them can afford. Low-interest loans from federal, state, and city urban renewal programs have made these houses especially enticing for young couples starting new households. An additional economic advantage is the proximity to work; the energy shortage and increasing transportation costs have made long-distance commuting from suburb to city a significant expense. Inner-city rehabs are often within walking distance to work, or at most a short bus or subway ride away.

The attractions of city life are not only economic, however— the chance to pick up a house at a bargain price—but also cultural and aesthetic. Being in touch with the artistic, sports, musical, and intellectual pulse of the city appeals to those who have no children and whose family ties are tenuous. Such fashionable sections as Society Hill in Philadelphia, Georgetown in Washington, New York's Brooklyn Heights, Chicago's Lincoln Park, Beacon Hill in Boston, the Garden District of New Orleans, and San Francisco's Cow Hollow attest to the lure of the inner city for those upper-income trendsetters who first saw the possibilities of charming, historic restorations in the heart of town. Less fashionable but equally well located neighborhoods offer the same advantages to middle-income couples willing to rehab on their own. Most young rehabbers choose houses with some architectural distinction: brownstones in New York, Victorians in San Francisco, federal homes in Washington. Much of the pleasure of homesteading comes not only from rehabilitating a house, but from the pride of restoring a whole block or neighborhood of nineteenth-century houses to their original splendor.

The aesthetic appeal of older homes is frequently cited as the motivation for rehabilitation. A Washington lawyer who rehabbed a row house in what was formerly an upper-income section of town pointed out that he has four fireplaces and solid oak floors, paneling, and molding, something he could never afford in a new house. A San Antonio couple noted that they were both well over six feet tall and they were attracted to the high ceilings and spaciousness of their older house. Another couple voiced a conviction held by many rehabbers: "We think older houses are just more interesting than new ones. You drive through a suburb and all you see are the same styles. Our house is different; it's both functional and unusual. No one in town has one like it."

Because rehabbing and homesteading have brought income back to the inner-city neighborhoods from which it has for years been draining, the movement is viewed as a social boon by city governments. But not by the blacks, ethnic minorities, the aged and poor who are displaced by the rehabbers. Black and ethnic minority leaders have cited the rehabilitation of inner-city neighborhoods as yet another example of the "trickle down" theory of urban renewal—benefits achieved by middle-income and professional-class rehabbers will *eventually* aid low-income families as well. Realistically, these critics argue, those benefits never come, because higher taxes and soaring real estate prices drive low-income families out of rehabilitated neighborhoods.

A 1978 HUD report challenges this view: a study of eighteen selected cities found that fewer than 100 or 200 households a year were displaced in most large cities. Donna E. Shalala, Assistant Secretary for Policy Development and Research, declared that "Far more poor people are being displaced today by housing abandonment and urban 'disinvestment' than by private restoration." And Anthony Downs of the Brookings Institute argues that some displacement is necessary and inevitable to reclaim cities. Preventing displacement, he declared, means either preserving cheap, substandard housing or providing costly housing allowances. "The fundamental problem of our cities," he said, "is that they've got too many poor people in them." Nevertheless, low-income families *are* being driven out of their neighborhoods by rehabbers—in Seattle, dislocations account for 19 percent of all moves—and there is scant satisfaction for those who are victims of inner-city rehabilitation to know they are only a small minority, or that they are unwitting pawns of sociological change.

The municipality that has experienced the greatest success in revitalizing inner-city neighborhoods through homesteading and rehabilitation—while still providing housing for low-income families—is Baltimore. This city, which was cited as the best example of urban renewal in America by the International Association of Planners, attacked the problem of inner-city decay by rescuing its ethnic neighborhoods, block by block, house by house. More than 500 abandoned homes have been sold to homesteaders, and the city itself has rehabilitated some 3,000 houses, selling half at bargain prices and renting the other half, all to low-income families. The projects have been financed by low-interest loans from the city. Some of the reasons why Baltimore has succeeded in an ambitious rehabilitation program whereas other cities have failed are:

The staff that administers the homesteading program is closely coordinated with other housing functions: planning, zoning, inspection, renewal, financing, and public housing. This close integration reduces bureaucratic inefficiency and the competition among city agencies.

The city has concentrated its efforts on entire neighborhoods rather than on piecemeal acquisitions.

Like the Self-Help Enterprises program, the homesteading staff closely screens applicants to ensure that they have the personal and financial capabilities to complete the rehabilitation work by becoming their own general contractors.

The staff makes a detailed cost estimate for each homesteader, supplies lists of approved contractors, checks on subcontractors' work, and provides technical advice when it is requested.

Because financing is provided by the city, loans are approved and administered quickly and efficiently.

James W. Hughes and Kenneth D. Bleakly, Jr., give detailed descriptions of the Baltimore program and other homesteading projects in *Urban Homesteading*.

A program that is this closely administered obviously reduces the builder's autonomy, but it also reduces the risk of failure and supplies a large measure of self-confidence: professionals are readily available for help when needed. However, because the program is focused on city blocks and neighborhoods, the informal networks of self-help builders that are so essential to any owner-builder enterprise still flourish. In some blocks of Baltimore virtually every house is occupied by a homesteader; help and advice are available next door.

One of the problems with HUD's urban homesteading is that, unlike Baltimore, many of the cities participating in it have eliminated the self-help component and have therefore turned it into simply another public housing program. The reasons cited for the insistence that all major work be contracted out to professionals in urban homesteading is that self-help building is an unreliable process that takes too long, sometimes doesn't get finished, and is difficult and costly to monitor. This has been disproven by Baltimore, Gary, Indianapolis, Dallas, and other cities that have encouraged self-help building. For example, the Dallas program provides technical assistance, training sessions, and even lends tools to homesteaders.

One suspects that there is more than arrogance involved in those cities which have eliminated sweat equity from HUD's homesteading program and deprived homesteaders of the chance to invest their own labor and creativity in a house. Pressures from contractors, the building trades, and unions no doubt influence all such municipal decisions. Indeed, the evolution of HUD's homesteading program has been in the direction of greater control of the entire process by the administering bureaucracy and less autonomy by the homesteader. Ultimately, this occurs because the program is heavily subsidized. When self-help builders finance their projects with loans or government grants they invariably lose a great deal of autonomy to the banker or bureaucrat. On the other hand, when homesteaders are spending their own money there is absolutely no reason why they should not be able to work on their homes any way they want to.

REHABILITATION AND RESTORATION

Such well-publicized renewal projects as Baltimore's are only the tip of the rehabilitation iceberg. As the cost of new construction soars, more and more Americans are finding older houses attractive as renovation and restoration projects. Harry J. Kane, executive vice president of Georgia Pacific Corporation, the world's largest plywood producer and a leading manufacturer of building materials, echoes a prediction made by numerous businessmen: renovating, remodeling, and rehabilitating will challenge home building in the 1980s. Recycling older houses makes environmental as well as economic sense, because every house that is rescued from demolition not only saves materials that would have gone into its replacement, but also helps retain the historic continuity that makes a city a living organism rather than a mere collection of buildings. A city that preserves its past is a dynamic, human place to live, and the people who participate in this preservation are some of the city's most valuable citizens.

Not every house is worth rehabilitating, of course; houses have a life span, like every other human artifact, and some are beyond saving. Brick and stone houses are better candidates

This dramatic contrast shows what rehabilitation can do for inner-city houses that have some architectural style. The house on the right looked like the one on the left before it was rehabbed. (Photo: HUD)

for rehabilitation because these materials sustain prolonged neglect with less damage. Wood frame houses must be checked much more carefully for structural damage. But many nine-teenth-century and early twentieth-century houses were built to last; if they have not been too severely mistreated, they can be successfully renovated. Would-be rehabbers who are looking for an older house to work on are usually aware that the worse off the house, the greater the economic leverage. Badly run-down houses can be purchased for a fraction of the cost per square foot of building a new house. But there are other eco-nomic considerations that can enter into a decision on whether or not to tackle renovation on a badly disintegrated house.

Whether one can live in the house while working on it is a prime consideration. If the house must be stripped down to a shell, as is the case with many inner-city houses, then rent or mortgage payments must be continued elsewhere during construction. This adds considerable expense to the project. If, on the other hand, the rehabber can live in the house while working on one room or one floor at a time, a considerable saving can be realized, and he or she is not under intense pressure to complete the work in order to move in.

One disadvantage of rehabbing room by room or floor by floor is that if interior walls and ceilings are to be replaced, as is often the case, plaster dust can be unpleasant and unhealthy to live with. It is virtually impossible to keep this dust from infiltrating every pore of house and householder. One way that some rehabbers cope with this problem is to seal off as much as possible of the part of the house being worked on until it is completed. Unfortunately, this is not always practical, es-pecially where plumbing and electrical renovations must be made, and these builders may decide to live elsewhere until the walls and ceilings are finished.

An advantage of living in the house during rehabilitation work is that by stretching out the process, it may be possible to complete the work without going into long-term debt. Those who rehab on their own have the option of attempting to com-plete the work entirely from income; this can sometimes only be done if they live in the house during renovation so they do not have to pay for the upkeep of two residences.

Those who build a house from the ground up enjoy the

benefit of starting with a *tabula rasa*; they are free to design their living space any way they like, with only the limits of budget and imagination to constrain them. Rehabbers, on the other hand, have no such liberty; they are bound by the decisions of the original builder. It is possible to enlarge rooms, make additions, redesign kitchens and baths, and give the exterior a face-lift; but major changes are usually done only at considerable expense. For example, whenever load-bearing walls are eliminated, engineering problems are encountered that may require the help of a professional. Many rehabbers seek the advice of an architect or engineer while still in the planning stage; homesteaders are sometimes offered such services at no charge by the city housing department.

Wiring and plumbing are especially ticklish problems for the rehabber. The electrical systems of all but the most recent houses are antiquated. Many older houses have 40 or 60 amp service, insufficient outlets, inadequate lighting, and worn or

Dilapidated, burned-out shells of houses can be given a new lease on life by innovative design. This inner-city house is getting new room additions and a contemporary roof line. (Photo: HUD)

CHIP: AN INNOVATIVE SELF-HELP REHABBING PROGRAM

One of the more innovative examples of self-help, low-income housing rehabilitation is the Chico Housing Improvement Program (CHIP) of Chico, California. CHIP is a nonprofit corporation founded in 1972. Their goal is to provide better housing for elderly and single-parent low-income families in its Sacramento Valley community. CHIP stresses self-help; it encourages its clients to take an active part in every step of the rehab process—planning, financing, and labor.

CHIP provides advice and packaging of loans, and counseling on such specifics as code inspections, materials, cost estimates, and energy conservation. CHIP supplies rehabbers with free student labor on rehabilitation projects—the most unusual part of the CHIP service. The students who work on CHIP rehab projects receive on-the-job training and course credit from California State University at Chico. The rehabber pays only for the materials.

The program is funded in part by the City of Chico, and by various private, state, and federal grants. The federal government's CETA program supplies an additional labor pool. Interestingly, approximately 30 percent of those working in nine crews on rehab projects are women interested in gaining experience in the building trades. CHIP is an outstanding example of how one community has succeeded in establishing a coalition of community elements—homeowners, university and city administrators, local business interests, and a variety of private and social agencies—in attacking the problem of substandard housing.

---

unsafe wiring. When a house is modernized it is almost always necessary to install a new service box with 150 or 200 amp capacity, to increase the number of outlets to the code minimum, and to do extensive rewiring. Plumbing is also often inadequate when you plan to add bathrooms and remodel the kitchen. Rather than attempt to make changes by cutting through existing interior walls, many rehabbers conclude that tearing out all plaster or dry wall and starting from scratch is the only rational way to proceed. Once this decision is made, however, the builder is committed to a major renovation job.

If estimating the amount of time a task will take is difficult for the owner-builder, it is almost impossible for the rehabber. Self-help builders are working entirely with their own measurements and new materials. They may make mistakes in laying out the house or in getting walls and partitions square and plumb, but they are in a position to correct them as they build. The rehabber has no such option. It is a rare older house that is anywhere close to being square or plumb; with age it has settled, shifted, and warped. Everything new must be custom-fitted to the inexactitudes of the existing structure. Dimensional lumber is often of different sizes, wiring and plumbing fittings are unconventional—nothing is standardized, as it is in new construction. How much time it will take to custom fit the new to the old often cannot be given even a ballpark guess. When it comes to refinishing old wood, sanding old floors, and stripping old wallpaper, the rehabber is in a no-man's land of time estimation. This is why rehabbers sometimes feel that they are restoring the pyramids.

In spite of the difficulties and the delays, however, rehabbing can be, and has been for many people every bit as enjoyable and creative as building a new house. There are many sources of self-fulfillment available to the rehabber that the new house builder does not have: the satisfaction of breathing new life into a house that has suffered neglect and abuse; the solution of design problems which transform a boxlike, constricted floor plan into an open, functional living space; the sensory pleasure of discovering quality materials, often beneath thick layers of paint—hardwoods, hand-carved moldings, fine stone or masonry work—and restoring them to their original beauty; the solidarity with neighbors who have also restored or rehabilitated houses in a formerly rundown city block and together have found in their mutual labors a sense of community where before there was isolation and distrust. No less than the owner-builder, the rehabber and restorer are also terpitects.

OWNER-BUILDERS
OF AMERICA:
A BLUEPRINT

Lewis Thomas, in his National Book Award-winning collection of essays, *Lives of a Cell* (New York: Viking, 1974), writes of the resemblance between social insects and humans. Ants, termites, and bees acting alone are little more than a collection of ganglia on feet, capable only of randomly moving from place to place. But when several of them gather together they seem to join their rudimentary cellular wits into motive and direction. And when hundreds or thousands reach a critical mass they can become, in Thomas's words, a "whole beast," which can be observed "thinking, planning, calculating."

In the workings of human social organizations something of the same kind of collective intelligence can be seen. The civil rights movement, the labor movement, the feminist movement, the Democratic or Republican parties—all appear at times to be no more than random collections of individuals pushing and shoving in head-to-head opposition, showing no clear purpose, no sense of rational action. But if you note the "beast" over a

199

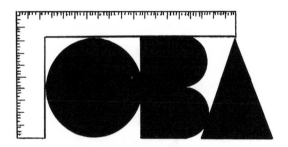

Owner-Builder's of America logo.

course of time, you see that it moves with deliberation and direction toward goals that can be measured, not by the inches of individuals, but by the miles of history.

Blacks, women, workers, Democrats, Republicans who are part of these movements live two lives: an individual life and a collective life. The individual life is accidental and personal; it follows the unique genius of "I." Its partnerships and bonds are familial, limited to a handful of individuals related by marriage, blood, or friendship. The collective life, however, throws out lines of communication far beyond the close tissues that unite family and friends. Like ants, wasps, and bees, the collective life bumps, pushes, and scrapes against other, often distant, members of the collective. The confusion of thousands of individual messages, seemingly like the Babel of Genesis, is decoded by the beast into one of those grand imperatives: equality, freedom, social progress.

THE SELF-HELP BUILDERS' COOPERATIVE

The owner-builder movement, although it possesses the necessary numbers of individuals, has never reached that critical mass where it acts as an intelligent, unified organism. In this respect, it is not a collective at all, but a *collection* of individuals. It is as if one termite, or only a handful of termites, produced a perfect abode for itself or its group, instead of masses of them that are essential to throw up what Thomas describes as "beautiful, curving, symmetrical arches and the crystalline architecture of vaulted chambers." Yet it is social and cooperative, and indeed could not exist as a movement at all if it were not. I have already described how networks of owner-builders interact in loose, informal exchanges of advice and information: lending tools, supplying labor, helping procure building materials, or simply offering encouragement or inspiration.

What is lacking is a collective voice that carries beyond local networks to community, state, and national audiences of owner-builders. If such an instrument could be found, if owner-builders were to cooperate at the neighborhood level, and if their voices were to join at the state and national levels, self-help builders could become a powerful, living force in deciding how this nation's citizens provide shelter for themselves. In contrast to the housing industry's billion-dollar message that the house is a purchased commodity, the collective voice of owner-builders could assert in organ tones that shelter is a creative artifact, a product of human craft that can be and is being built by its owners.

I firmly believe that such a collective is a possibility, but if it is to articulate its voice it must be totally independent—particularly of any kind of government. It must be a cooperative of individuals whose only allegiance is to the self-help idea; and it must be organized, controlled, and administered by owner-builders. This is not to say that it would not seek to influence government action in behalf of the self-help builder, but it would remain steadfastly private and nonprofit. The message of such a collective voice would be monolithically simple: to promote and encourage the construction of houses by their owners.

I have given some thought to the matter and would like to set forth some ideas for a national cooperative of owner-builders which I will conditionally call Owner-Builders of America (OBA). In my questionnaire to self-help builders I asked, "Would you be interested in belonging to an owner-builder cooperative association that would help others in your community build their own houses, if such an organization existed?" An overwhelming majority responded affirmatively; some declared that by helping family members and friends, an informal cooperative already existed in their communities. This response has helped convince me that there is a readiness among owner-builders for a community cooperative effort beyond that which already exists in the informal networks that every self-help builder taps into and helps sustain.

The willingness to help others build their own homes derives, I believe, from three motivations. These would be the driving force behind an owner-builder cooperative movement.

1. The desire to share with others an exciting, fulfilling, even passionate creative interest. This is the basic motivation behind all artistic, hobby, and special interest groups.
2. The need to continue to put to use the newly-acquired managerial and construction skills that a person acquired in the course of building a house.
3. A sense of community that is expressed in an altruistic desire to help others.

Few owner-builders would be driven by all of these impulses, but only one is necessary to provide the stimulus for joining a self-help builders' cooperative. The foundation of the organization would be the neighborhood group, for it is at this level that all significant action would occur; it is here that self-help builders would encourage and aid others to build their own houses. The membership of the neighborhood organization would consist mainly of two groups: those who have already built houses or are in the process of building them and those who have the desire to build. Other interested community members might also be included, but they would represent a distinct minority. The main criterion for membership would be a desire to encourage and promote self-help building. There would be no restrictions as to age, sex, race, profession, or income; in fact, the more heterogeneous the membership the greater its potential for growth and community impact. There would, of course, have to be dues, officers, and meetings, but the organizational apparatus would merely exist as an armature to support the real work of the organization: helping owners build their houses. The resources of the neighborhood organization would fall roughly into the following categories: information, education, tools, labor, and community action. I discuss each of these separately.

INFORMATION

One of the most valuable resources of a neighborhood Owner-Builders of America organization would be its data bank. This information library would include a collection of materials, contributed or loaned by its members, on every phase of planning, financing, construction, and management of a house. These materials would include books, magazines, pamphlets, and manuals on construction; house plan books, blueprints, suppliers' catalogues; printed material of any kind on financing, money management, and energy efficiency. Up-to-date information on local money supply, real estate markets, and housing conditions would be particularly useful for new members of the community.

The primary source of information, however, would be the owner-builders who have already completed their own houses. A card file would be kept on all members who have completed houses, listing the kinds of houses they have built, photographs, and any significant construction details that might be useful to other builders. Owner-built houses themselves, either completed or under construction, would be another useful resource. These would serve as models for the kinds of work that other members are contemplating, planning, or constructing, and also serve as motivation and inspiration for would-be builders. The data bank would, in effect, formalize and make more readily available the informal networks that self-help builders manage to establish in the course of their construction enterprise.

EDUCATION

An Owner-Builders of America organization could become the hub of a neighborhood house-building education program. Many communities throughout the nation already have technical education centers, vocational schools, university extension programs, and high schools offering courses in house building. But these present only a fraction of the potential. Most of the courses now being offered have come about because of the energy, vision, and dedication of a single individual, usually a terpitect. There are few organized, community-wide programs. The most effective educational programs are those which not only teach individual construction skills, such as carpentry, masonry, or plumbing, but which organize students into teams that actually build or rehabilitate a house. Such programs teach house construction from the foundation up in the most effective way it can be done—by giving students the chance to work on each building system in a real situation. An OBA organization could become a resource center for vocational house-building courses; it could provide not only information from its data bank, but also volunteer teachers. Retired owner-builders could possibly devote a considerable amount of time to supervision and training.

It might also be possible for OBA members to organize and operate their own house-building training program if there is

A housebuilding class at The Owner Builder Center in northern California. An owner-builder cooperative could encourage students like these to learn construction skills by on-the-job training. (Photo: Herb Ziegler)

no vocational or extension school nearby. Again, it could be staffed by elderly or retired members, and part-time teaching and training could be provided by members of all ages. Private house-building training centers such as the highly successful Shelter Institute, and Cornerstones, both in Maine, and The Owner Builder Center in California have demonstrated that there is a widespread demand for adult owner-builder training. Programs like the Self-Help Enterprises training plan could serve as a model for neighborhood OBA educational efforts.

It would not be necessary, of course, for formal courses or training programs to be initiated. Any time OBA members help another member build a house, the opportunity for on-the-job training presents itself. The educational function of the organization would best be served by satisfying community needs in the simplest, most direct way. In most cases, this is accomplished by working closely with existing educational institutions.

Another educational function of the neighborhood organization would be to enlighten the community about the personal benefits and creative joys of terpitecture. This is essentially a public relations task and could be approached in a number of ways. Self-help building is a human interest story, one that has appeared in literally thousands of newspapers and magazines in every part of the country. It is not difficult to publicize owner-builder activity because the public is intensely curious about it. As I have already pointed out, the idea of building your own shelter strikes a deeply felt chord in all of us. Publicizing self-help building in the communication media—newspapers, radio, and TV—could become one of the most effective ways of educating the public about owner-building. Other methods of disseminating information about OBA and its work might be organizing an annual tour of owner-built homes in the community, holding public seminars on self-help building and OBA, and inviting speakers to public forums on owner-building.

TOOLS

One of the most useful services Owner-Builders of America could provide its members would be a tool bank. Members would donate any duplicate or surplus tools they may have or would agree to lend construction tools and equipment to other members. An up-to-date card file would list those tools a member was willing to lend and the terms of the loan.

Lending tools presents a special problem that every owner-builder is aware of—it's a rare worker who likes to lend tools. They are personal extensions of the owner and, through use, tools have become warm friends. The craftsman knows their idiosyncracies, their capabilities, and their flaws. A builder also knows that when tools have been loaned in the past they have been returned late, broken, dirty, and dull. If a tool bank is to be workable it must ensure that both borrower and lender are satisfied with the transaction. Borrowers must be made aware of their responsibilities, and the lender assured that the risks to tools and equipment are minimal. A few simple rules ought to satisfy both conditions:

All tools and equipment should be returned in *better* condition than when they were borrowed—clean, well-oiled, full of gas where applicable, and in perfect working shape.

Any tool broken, completely or in part, after it is borrowed, must be repaired before it is returned. This is not only a courtesy to the lender, it ensures that the tool will be there for the next borrower. Lost tools must be replaced, by the borrower.

Blades for cutting tools must be sharpened or replaced before the tools are returned.

All tools and equipment borrowed for a limited time should be returned promptly.

Owners must be able to recall any tools or equipment at any time for their own use.

If these rules were observed, a tool bank should function with a minimum of complaints from lenders and borrowers. The system would probably work most effectively if tools were loaned through a third, mediating, party. This mediator would keep records and make sure the borrowing rules are observed. A lender would not have to confront a borrower with complaints about the condition of tools or equipment.

LABOR

A labor bank, like the data and tool banks, would mobilize for the owner-builder a necessary resource. This system would use two card files. The first card file would list the skills of all OBA members and their willingness (or unwillingness) to volunteer work on a fellow member's house. In my opinion, most members would be willing to contribute free labor to other owner-builders. I asked my sample of owner-builders whether they had helped another self-help builder. A majority of those responding said they had, and many had helped more than once. This willingness to donate time, labor, and advice to other house builders has been, and continues to be, one of the striking features of the owner-builder movement. It is not only self-help; it is mutual help, and therein lies one of its great strengths and its potential for growth and influence. A family that has received voluntary labor will be encouraged to contribute its own labor to others. Owner-Builders of America would act as a facilitator of this cooperative self-help. In some communities, groups of families might be encouraged to join in a mutual building team, like the Self-Help Enterprises groups, and build two, three, or four houses simultaneously, each family working on all of the houses.

The kind of work that members would contribute to the labor bank, at what times, and how much, would have to remain extremely flexible. Perhaps a self-help builder could simply announce that he or she is putting up dry wall next Saturday and distribute a sign-up sheet. The ideal model for such voluntary labor would certainly have to be the old-fashioned barn raising, a practice that is still observed by the Mennonites. In the course of a single weekend, dozens of families from surrounding communities converge on a farmer's building site where a foundation is ready and, working like one of Lewis Thomas's insect colonies, they throw up a magnificent barn. It is supervised by elders who have been building such structures for generations and could assemble one blindfolded. The occasion is a huge party with vast quantities of food and fellowship.

Sharing work on houses for the members of a building team would be one of the goals of Owner-Builders of America. (Photo: George Ballis)

I see no reason why an Owner-Builders of America chapter could not have a framing party and completely frame a member's house over a long weekend. Or a roofing party. Or a siding party. Members and their families could bring their own tools and equipment and the self-help builder could supply food and refreshments as his or her contribution to the event. A skilled member could supervise the work, and whole families could join in. Such cooperative work efforts would produce a solidarity among members that no amount of meeting and talking and debating—the normal activities of most organizations— could hope to match.

The second card file in the labor bank would be an annotated reference list of professional subcontractors, craftsmen, and skilled and unskilled workers in the area. OBA members who had hired professionals would evaluate their performance for the card file. Reports from other reliable sources could also be included. Care would have to be taken that such a file does not become a blacklist of certain subcontractors and construction workers. A rating system could be devised that would score professional workers on such items as fairness of fees, work quality, promptness, dependability, and honesty. A reference tool like this could give the owner-builder some objective standard, other than hearsay or ignorance, for choosing subcontractors or construction workers. If professionals in the area knew they were being evaluated, the reference file could also produce subtle pressures to improve work quality and standards—at least where OBA members were concerned.

COMMUNITY ACTION

A strong Owner-Builders of America organization would not only seek to satisfy the individual housing needs of its members; it could also be a political force in its community. A legitimate aim of OBA would be to become actively involved in all community affairs that touch on housing. In one respect, it would function as a homeowner's association; indeed, where the goals of OBA coincided with those of local homeowners' groups, it could form alliances with them. OBA would be interested in such issues as zoning, land use, urban planning and development, city services, highway relocation, school location and funding, housing subsidies, and property taxation. But more particularly, it would seek to liberalize building and occupancy codes and to alert lending institutions to the existence of self-help building as a credible alternative to commercial contracting.

In short, OBA would function as a pressure or lobbying group in those areas of community life that involve its members as both homeowners and house builders. Its effectiveness as an action group would depend largely on the size of membership, the commitment of members to promoting the organization's aims, and the existence of emotionally-charged coalescing issues. These elements would obviously vary widely from group to group.

NATIONAL AND REGIONAL ORGANIZATIONS

The parent Owner-Builders of America organization would tie the community groups into a unified, nation-wide informational network. Such an organization would function in two modes. In the first mode it would act as a central data bank where useful information from community groups would be distributed to member groups and individuals. In the second mode it would function as an advocate for the owner-builder movement in this country. In this capacity, it would use the strength derived from a national constituency to facilitate, promote, and encourage self-help housing in every possible way.

As the organization grew, city, state, or regional satellites might be desirable to promote the interests of particular areas. There is no question that the owner-builder in the Southwest, for example, builds with different materials and under different spatial, geographical, climatic, and political conditions than the builder in the Northeast. Those who reside in rural areas do not have the same problems as those who are building in the suburbs or inner city. Regional organizations could address themselves to the particular needs of their city, state, or region and could arrange local meetings or conventions conveniently and economically. Regional organizations would limit themselves to issues and interests of their particular areas, and would represent those areas and bring those concerns to the attention of the national body.

The voice of the national OBA would be a publication distributed to all members; it would be financed by annual dues and perhaps by paid advertising. (The question of whether advertising would compromise the independence of the organization would have to be determined by its membership.) Through the OBA publication, members could tap into a national network of owner-builders. This single instrument would make it possible for individuals and groups from every part of the country to share construction experience, technical know-how, financing and organizational information, mutual aid techniques, energy-saving methods, new products information, and news of government action affecting the owner-builder.

The publication could include feature articles on unusual owner-built houses, on projects that might be copied by other locals, on ways individuals have managed codes and regulations, on financing strategies, on creative solutions to building problems, on new construction materials and techniques of particular interest to the self-help builder, on membership recruiting ideas, and on the organization's activities as owner-builder advocate. Regular departments would include reviews of construction manuals, how-to literature, and other subjects related to self-help building; reports on package or precut house-building experiences; announcements of local meetings, conventions, and seminars; and a letters forum where members could voice opinions and complaints.

One further goal of the OBA publication would be to disseminate information about owner-building to the public at large—a public relations function. This could be accomplished by distributing OBA publication articles to the mass media to ensure a nationwide audience. In publicizing OBA, it would be essential to let the public know that film and TV stars, sports figures, businessmen, doctors, lawyers, and clergymen also build their own houses; and that saving money is not the only, or even the most important idea behind self-help building. Owner-Builders of America, by publicizing self-built houses of every kind—from small retirement homes to rehabs to mansions—would demonstrate that building a house is within the realm of possibility for nearly everyone, that it is truly a democratic and democratizing endeavor. One of the most valuable services a national OBA organization could perform would be to bring together under a single roof, as it were, potential terpitects of all kinds—men and women; poor and affluent; white collar and blue collar; black and white; young and old.

The national OBA organization would also take an active role in encouraging and promoting house-building courses in the nation's educational institutions. This could be accomplished in a number of ways. Articles in the OBA publication would report on successful vocational education groups to encourage other institutions to offer self-help home construction courses. OBA could also offer its cooperation in the production of textbooks, manuals, and handbooks for educational use.

Owner-Builders of America would promote and encourage housebuilding courses such as these students of the Owner Builder Center are participating in. Teaching by doing is the classic way of passing on housebuilding skills. (Photo: Owner Builder Center)

OBA could play a role in promoting and reporting on research into self-help building methods and techniques. The housing industry designs houses for professional construction; owner-builders must adapt to these professional systems by learning each "trade" as they progress. New building systems adapted to the needs of the self-help builder—unskilled or partially-skilled—could be explored with OBA aid. The organization could sponsor a national competition for houses designed specifically for owner-building, stressing efficiency, economy, and minimal skills. Professional architects and architectural schools, as well as OBA members, could be encouraged to participate in designing an "OBA model house." Plan and blueprints of winning designs could be made available to OBA members at cost, and completed houses from the designs could be publicized nationally. A competition such as this would not only encourage research into self-help design but would also promote interest and membership in OBA. Other research efforts could be directed toward the use of solar energy and energy-efficient techniques in self-help construction, and toward innovative approaches to such mutual self-help projects as cooperative construction of apartments or condominiums.

I have stated that it is essential that a national owner-builder association remain independent of government control and influence. This does not mean, however, that an autonomous, democratically-supported, nonprofit OBA would not throw all its energy and resources into influencing government action in those areas where government comes into contact with individual home construction. As advocate of the owner-builder movement in this country, OBA would function as a lobbying group not unlike other national consumer organizations. In establishing an advocacy position, OBA would be faced with an inherent philosophical conflict between its own goals and the beliefs of many business and government leaders. The individual who builds his or her own house cannot help but inspire the admiration of the most conservative members of the business and government communities. The owner-builder acts with individual autonomy, self-reliance, independence, entrepreneurship. He or she is dedicated to thrift and hard work, traits which have traditionally been associated with the free enterprise system. In self-help building, however, these characteristics do not produce a profit for realtors, commercial builders, or developers; on the contrary, they are in opposition to the philosophy of consumerism which is the backbone of the housing industry. Self-help builders are not customers for a house; therefore, their building activity is subversive to the housing industry. Mutual self-help, where groups of families join together to build several houses, even smacks of socialism.

This conflict is perhaps most visible in the case of low-income self-help housing projects and proposals. The housing industry would like to solve the problem of low-income housing as it always has—with massive housing projects which produce substantial profits for everyone concerned in the industry. To have low-income families build their own houses produces nothing but—well, houses for low-income families. It does not produce housing industry profits nor, it might be added, does it produce union wages. So it is not surprising that the owner-builder will fail to find any support among those associated with the construction unions, with the housing industry, or among the industry's numerous friends in the Department of Housing and Urban Development, which draws many of its high echelon officials from the ranks of the industry.

Even though HUD officials may pay lip service to the philosophical goals of the owner-builder (who is, after all, an independent businessman freely conducting his or her own enterprise) they are not likely to offer any aid. Unlike the housing industry, owner-builders today have absolutely no political clout. Not until their voices are heard, not until they become a political entity, will owner-builders receive more than token recognition from Washington. A primary task of a national OBA organization would be to apply such pressure. It may not necessarily concentrate on HUD, whose bureaucracy is a quicksand of inertia where the individual owner-builder is concerned, but focus directly on influencing Congress. One of the best hopes for owner-builders in America would be for OBA to enlist the active support of those members of Congress whose parents or family members have built homes or who have been owner-builders themselves. A handful of powerful congressional owner-builders, acting in unison, could do more for self-help building in this country than any amount of direct HUD lobbying.

The priorities of an OBA advocacy program would necessarily be flexible, but a high priority would certainly be given to the adoption of standardized national building and occupancy codes which recognize self-help building as a separate construction category. OBA could draw up a model building code incorporating special classifications for owner-built or owner-rehabilitated dwellings. These new classifications would waive many of the requirements that apply to houses built for profit by contractors and developers.

OBA would also want to ensure that all government loan subsidies that become available to contractors and developers are also made available to individual owner-builders and mutual self-help groups. Historically, government agencies have been reluctant, even where legislation mandates it, to become involved in processing individual owner-builder loans. In Turner's

Freedom to Build, Richard B. Spohn observes that HUD's programs are structured to serve the large-scale projects of the housing industry rather than single self-help builders, and that HUD is geared to process commitments in blocks of hundreds rather than handle loans to small, mutual self-help associations or to individual owner-builders. OBA would lobby for special individuals or agencies within HUD and the Department of Agriculture to administer subsidized loans for the single owner-builder. OBA could also insist that all subsidized low-income housing programs, such as urban homesteading, include a self-help option, building and occupancy code waivers, and technical assistance.

A final idea OBA could explore is a direct tax credit to the owner-builder who completes a house without a loan. The government presently subsidizes homeowners by allowing them to deduct the interest paid on home mortgages from their income taxes. The individual builder who constructs his or her house without a loan, however, is rewarded for industry and economy by losing this subsidy. Yet this method of building is one of the most cost-efficient kinds of home construction; it produces a high equity-to-income ratio, adds to the housing store, and costs the public nothing. If the government is to continue subsidizing mortgaged home construction—and, in effect, bank and savings and loan profits—it is only equitable that the owner-builder who chooses to complete his or her house without a loan should receive a comparable subsidy in the form of tax credit. Such credit would encourage the most efficient and economical kind of construction ever devised— houses built by their owners and paid for with cash.

THE NATIONAL CONVENTION

All of these OBA activities would come into focus at an annual convention, a national meeting of individuals from community locals and from city, state, or regional associations. The annual convention would be an opportunity for a personal exchange of information, viewpoints, and opinions; for displaying creative and innovative approaches to owner-building; for planning future OBA strategies; and for viewing new construction products and materials of particular interest to self-help builders. Seminars, symposiums, and colloquia could be held on such diverse topics as solar energy, insulation, skills, package houses, community action, low-income construction, mutual self-help projects, financing, and self-help research. The meeting would be, in effect, a national forum for all of the various interests and activities of the organization's members, and would not only generate excitement for self-help building within the organization, but would also help publicize the work of OBA in the media.

BUDGETING

The program I have outlined for Owner-Builders of America is ambitious and would require a significant budget—if not at the local level, then certainly for the national organization. The locals could, I believe, be funded entirely by modest membership dues. Space for OBA's local headquarters could be sought at no cost from local community organizations—a church basement, or rooms in the civic center, library, or school—and staffing would be entirely voluntary. However, the national organization would require some permanent professional staff if it were to accomplish even a portion of those goals I have suggested. Where would this money come from?

OBA publications—magazine, manuals, handbooks, blueprints—could produce some revenue, and dues would supplement it. I believe, though, that additional financial support could be obtained from foundations or institutions whose self interests are aligned with those of the owner-builder. In particular, the manufacturers of building materials and the associations and foundations they support would be likely sources of financial backing. Private foundations whose philosophies are in agreement with the creative and humanistic goals of OBA would also be solicited. One of the initial tasks of the nascent OBA would be to seek such support. Unlike government-funded organizations, OBA would not, I think, be top-heavy with professional staff. Indeed, it should rely on a voluntary staff as much as possible. Leanness, efficiency, and economy—traits which are the guiding principles of self-help builders' personal enterprises—should also be a characteristic of their cooperative endeavors.

OBA AND THE ELDERLY

One of the sources of voluntary staffing at all levels of the organization could be the elderly. I was frankly quite surprised when my sample of owner-builders disclosed that a significant number were retired men and women in their sixties and seventies. Thousands of retired couples have taken as a retirement project the construction of a new home. And it is a marvelous project for the elderly; it is richly creative. Retired men and women have the time to devote all of their energies to it, and they can often finance the project entirely from the sale of a larger house they no longer require.

OBA neighborhood organizations could take an active part in aiding the elderly in their self-help construction projects. Those older members who are physically unable to do the labor involved could draw from the labor and skills pool of the local membership. The elderly, in turn, could supply much of the staffing requirements of the organization. Helping elderly, low-income couples build their homes would be one of the more rewarding projects of an OBA local. Mutual self-help retirement houses or apartments could also become a vital part of the organization's work.

In the next two decades, as the number of over-sixty citizens increases dramatically in this country, housing the elderly will become as much of a social problem as housing the poor is today. OBA could play a significant role in aiding elderly individuals and couples living only on Social Security to create shelter for themselves. Research into the problems of the elderly has repeatedly demonstrated that inactivity, creative stagnation, lack of significant goals, and poor social interest account for many of the debilities of aging even more than declining energy and poor health. Encouraging the elderly to become terpitects—not only to throw their energies into constructing their own houses, but to participate in community efforts to aid other elderly citizens house themselves—could be exciting and contagious.

THE SOCIAL IMPACT OF OBA

The organization I have sketched above, or one something like it, is the only way I can see that the presently scattered, disjointed individual efforts of owner-builders can be unified into a cohesive national movement. There are those who would argue that there is no need for such joint, cooperative endeavor, that the self-built house is a romantic throwback to a handicraft era which is best abandoned to the efficiencies and economies of the mass-manufactured house. Those who see no future in the owner-builder movement claim that it will never solve a housing problem; it will not produce low-income houses for the poor, for minorities, or for the elderly, and it will never rise above its present position in the housing picture as an insignificant, middle-class "hobby" activity of no social importance.

I agree that an owner-builder association, no matter how strong and effective, could never solve a "housing problem," particularly if the problem is defined as finding adequate housing for the poor, elderly, and disadvantaged. A cooperative owner-builder movement could do little about rental housing or about house ownership for those who are incapable or unwilling to assume the responsibilities of constructing their own shelter. These two categories admittedly include a large percentage of families who constitute the nation's "housing problem." There are, however, a large number of poor, elderly, and disadvantaged families who are quite capable of helping themselves if they knew how to do it, if financial aid and education were available, if a support organization existed to ease them through the process. For these families, a self-help builder's association could be the difference between inadequate housing and a home of their own. If an owner-builder's cooperative organization succeeded in encouraging and aiding only a handful of such families to become homeowners, it would contribute to the solution of the nation's housing problem. I believe the potential for this kind of aid and encouragement could involve hundreds of thousands of families.

To the criticism that self-built housing in this country is merely a "hobby" for the middle- and upper-income family and that an owner-builder cooperative organization would be no more socially significant than a car customizing club or an association of aircraft or boat builders, I would respond this way: the American man or woman who builds a house is part of a cultural phenomenon that transcends "hobby" or "avocation" because it is an expression of human creativity as it is applied to territoriality and shelter. Whether the need to imprint one's brand on a piece of space is a loop in our genetic circuitry, as sociobiologists suggest, or merely a deeply-rooted cultural tradition, there is ample evidence that building houses is one of the most richly satisfying forms of creative behavior. Because of this, the owner-builder phenomenon more properly belongs to the self-actualizing experiences of Abraham Maslow and the human potential movement than to the sociometric equations of the housing specialist. To this kind of building experience I have given the name terpitecture; Owner-Builders of America could be an association of, by, and for terpitects.

The author's home at Clemson, South Carolina.

BIBLIOGRAPHY

The literature on self-help home construction is so vast and grows so rapidly that those interested either in owner-building or in exploring the owner-builder movement are likely to be confused by the sheer quantity of titles displayed at a well-stocked bookstore. The following list offers a sampling of what I and other owner-builders have found to be some of the most useful books. Readers would be advised to borrow as many titles as possible from a library before purchasing them.

Because prices are so volatile they have been omitted, but many of the titles are in less costly paperback editions. Government publications are always a bargain. Prices for these can be obtained by writing the Superintendent of Documents, U.S. Government Printing Office, Washington, DC 20402.

Anderson, Bruce, with Michael Riordin. *The Solar Home Book*. Harrisville, NH: Cheshire Books, 1977.
One of the more popular books on heating, cooling, and designing a house with the sun in mind. Unlike many books on solar energy, this one has a do-it-yourself bias.

Blackburn, Graham. *Illustrated Housebuilding*. Woodstock, NY: The Overlook Press, 1974.
A decidedly counterculture description of house construction. It is simple to read, and its illustrations are fittingly handmade. This is one of many books on owner-builder construction by small, antiestablishment presses; it captures the grass-roots approach to the entire movement.

Boericke, Art, and Barry Shapiro. *The Craftsman Builder*. New York: Simon & Schuster, 1977.

———. *Handmade Houses*. New York: A & W Publishers, 1973.
These books are for inspiration rather than information. They demonstrate that craftsmanship and love of materials have nothing to do with economics.

Brand, Stewart, ed. *The Next Whole Earth Catalog*. New York: Random House, 1980.
This is the latest edition of the National Book Award-winning Whole Earth Catalog. *It is one of the best places to go shopping for literature about the owner-builder experience.*

Bright, James. *The Home Repair Book.* New York: Doubleday, 1977.
A large-format loose-leaf binder with sections on most house systems as well as on tools and materials. The illustrations are clear and easy to follow.

Browne, Dan. *The Housebuilding Book.* New York: McGraw-Hill, 1974.
Weak on illustrations, but Browne is a builder and teacher who is extremely knowledgeable about all house systems.

Ching, Francis D. K. *Building Construction Illustrated.* New York: Van Nostrand Reinhold, 1975.
This is a large-format book, richly illustrated with line drawings of the systems of a house and particularly good on framing systems.

Coffee, Frank. *The Complete Kit House Catalog.* New York: Pocket Books, 1979.
Most package house manufacturers in this country are represented here by a single model of their line of homes. A floor plan and description of the house is included. Information is from the manufacturers and is not critical. It is one of several books of its kind.

Cole, John and Charles Wing. *From the Ground Up.* Boston: Atlantic-Little, Brown, 1976.
In 1974, both authors were instrumental in forming The Shelter Institute, one of the most publicized house-building schools in the country. Each has built his own home so they are knowledgeable on the subject. Wing now teaches home-building at Cornerstones, a school for owner-builders in Brunswick, Maine.
This book is not a construction manual per se. It deals with theory and engineering details and is more concerned with the design and planning of a house than with actual construction. There are excellent sections on site planning, tools, and engineering how-to information. It is oriented toward the New England climate.

Concrete and Masonry. U.S. Government Printing Office, 1970. S/N 008-020-00313-1.
A U.S. Army manual that covers the subject with military thoroughness.

Conran, Terence. *The Bed and Bath Book.* New York: Crown, 1974.
————. *The House Book.* New York: Crown, 1974.
————. *The Kitchen Book.* New York: Crown, 1974.
These books are expensive full-color picture books that are a delight to browse through for ideas, but are essentially coffee-table books. Fun to look through if you can pick them up at the library.

Daniels, M. E. *Fireplaces and Wood Stoves.* New York: Bobbs-Merrill, 1977.
A handbook on the construction of fireplaces and the installation of prefab fireplaces and wood stoves.

DeCristoforo, R. J. *DeCristoforo's House Building Illustrated.* New York: Harper & Row, 1978.
This is an exhaustive guide to house building by a professional writer in the field. Illustrations are good and plentiful—over 1,000 of them. Many libraries have this book.

DiDonno, Lupe, and Phyllis Sperling. *How to Design & Build Your Own House*. New York: Knopf, 1978.
One of the best single-volume books on house building from the foundation up, by two women architects. It is especially good on design, the authors' forte. Illustrations are all line drawings, clear and legible.

The First Passive Solar Home Awards. U.S. Government Printing Office, 1979. S/N 023-000-00517-4.
A catalog of award-winning passive solar designs. A marvelous idea book.

Haney, Robert; David Ballentine; and Jonathan Elliott. *Woodstock Handmade Houses*. New York: Random House, 1974.
Like The Craftsman Builder, *this book is for inspiration. It contains many color photographs of alternate life style owner-built houses.*

Hard, Roger. *Build Your Own Low-Cost Log Home*. Charlotte, VT: Garden Way, 1977.
A book that covers log building systems, both from scratch and with kits. A well-illustrated overview.

Home Repair and Improvement. Alexandria, VA: Time-Life Books, 1977.
This is a series of large-format books on various construction systems. Series titles include: Kitchens and Bathrooms, Outdoor Structures, Paint and Wallpaper, Floors and Stairways, Roofs and Siding, New Living Spaces. *The layout is excellent and the illustrations are exceptionally clear and easy to follow. These are sold as a series through the mail but can be purchased individually in bookstores. Many libraries have these books.*

Hotton, Peter, *So You Want to Build a House*. Boston: Little, Brown, 1976.
Hotton is Home and Garden editor of the Boston Globe *and writes in a clear, journalistic style. It is a short manual, full of useful information, but the illustrations are minimal.*

Hughes, James W., and Kenneth D. Bleakly, Jr. *Urban Homesteading*. New Brunswick, NJ: The Center for Urban Policy Research, Rutgers University, 1975.
This is an informative, well-researched treatment of the homesteading phenomenon.

Kern, Ken. *The Owner-Built Home*. New York: Scribner's, 1975.
For years, Kern has been a leader in the owner-builder movement. His book has been something of a bible for counterculture home builders. Although some of his ideas may not be relevant for middle-class builders, every would-be owner-builder will find him fascinating. Illustrations are funky but clear.

Kern, Ken, with Ted Kogan and Rob Thallon. *The Owner-Builder and the Code*. Oakhurst, CA: Owner-Builder Publications, 1976.
One of the best things written about the personal and social problems of coping with codes. This also follows a counterculture approach.

Mazria, Edward. *The Passive Solar Energy Book*. Emmaus, PA: Rodale Press, 1979.
A comprehensive sourcebook by an architect who has been highly influential

in promoting passive solar energy. A primer, technical manual, and work-book containing a wealth of information on design of passive solar systems. The U.S. Department of Energy has adopted Mazria's criteria for passive solar design, making his book a standard reference work in the field.

Neal, Charles D. *Do It Yourself Housebuilding Step by Step.* New York: Macmillan, 1973.

Somewhat limited in the kind of house described, but the construction of basic house systems is clearly illustrated.

Nearing, Helen and Scott. *Living the Good Life.* New York: Schocken Books, 1954, 1970.

The Nearings were two of the first owner-builders to write about the joys of subsistence living and building. They are especially informative about building with rock.

Reader's Digest Complete Do-it-yourself Manual. Pleasantville, NY: Reader's Digest Association, 1973.

Houses have been completely rehabbed with the help of this single book. An encyclopedia of house repair and remodeling. It ranks with the Time-Life series for the excellence of its illustrations. Expensive, but many libraries have it.

Reducing Home Building Costs with Optimum Value Engineered Design and Construction. Rockville, MD: NAHB Research Foundation, 1977. (626 Southlawn Lane, Rockville, MD 20850.)

A first-rate, inexpensive guide on how to cut construction costs without sacrificing safety. It covers all house systems, and is, in effect, a building manual. A must for all owner-builders.

River. *Dwelling.* Albion, CA: Freestone Publishing Co., 1974.

Dwelling *represents the counterculture movement in California. It is a collage of photographs, line drawings, poems, and impressionistic prose— all variations on a theme: the joy of building. I think it is one of the loveliest books the owner-builder movement has produced.*

Scharff, Robert. *The Complete Book of Home Remodeling.* New York: McGraw-Hill, 1975.

A one-volume book on rehabbing and renovation. Good sections on kitchen and bathroom remodeling, and on redoing plumbing and heating systems. Clear illustrations.

Solar Dwelling Design Concepts. U.S. Government Printing Office, 1976. S/N 023-000-00334-1.

A clearly written introduction to solar design.

The Solar Energy Handbook. Popular Science Magazine.

Sunset Homeowner's Guide to Solar Heating. Menlo Park, CA: Lane Publishing, 1978.

Several hundred thousand copies of this low-cost, well-written, well-illustrated introduction to the subject are in print.

Sunset Building and Remodeling Books. Menlo Park, CA: Lane Publishing.

This is a series of inexpensive, large-format paperbacks put out by Sunset *magazine. They include individual books on such subjects as plumbing, carpentry, wiring, painting, and wallpapering, decks, fireplaces, patios, bookshelves and cabinets, kitchens, bathrooms, tile, and house planning. They are mostly idea books with numerous black-and-white photos and varying amounts of construction information. The orientation is West coast, ranch-style, and contemporary.*

Turner, John F. C. *Housing by People.* New York: Pantheon, 1976.
This is an excellent theoretical work on the owner-builder phenomenon.

Turner, John F. C., and Robert Fichter, eds. *Freedom to Build.* New York: Macmillan, 1972.
A collection of essays on the economics, politics, and sociology of the owner-builder movement throughout the world. The best book of its kind in print.

The Underground Space Center, University of Minnesota. *Earth Sheltered Housing Design.* New York: Van Nostrand Reinhold, 1978.
This is a comprehensive treatment of the underground house—a summary of current research on building underground or earth-berm houses, showing design and engineering considerations. Very well researched.

Vivian, John. *Wood Heat, new and improved edition.* Emmaus, PA: Rodale, 1978.
From woodlot to fireplace, from stove to flue—an exhaustive treatment of the subject.

Wade, Alex. *A Design and Construction Handbook for Energy-Saving Houses.* Emmaus, PA: Rodale, 1980.
Wade's previous book, 30 Energy-Efficient Houses You Can Build, *has well over 100,000 copies in print, and this one probably will be equally popular. It contains specifics on financing, designing, and building thermally efficient houses.*

Wagner, Willis. *Modern Carpentry.* South Holland, IL: Goodheart-Willcox, 1979.
This is usually considered to be the standard text for carpentry.

Wampler, Jan. *All Their Own.* New York: Oxford University Press, 1978.
All Their Own *is a delightful collection of folk art houses—whimsical, crazy homes made of every kind of material imaginable.*

Watson, Donald. *Designing & Building a Solar Home.* Charlotte, VT: Garden Way, 1977.
A popular handbook; well over 100,000 copies in print.

Weiss, Jeffrey. *Lofts.* New York: Norton, 1979.
A color picture book of ideas, with little construction advice, but it shows what imaginative builders have done to turn a loft into an attractive, livable space.

Wells, Malcolm, and Irwin Spetgang. *How to Buy Solar Heating Without Getting Burnt.* Emmaus, PA: Rodale Books, 1978.

A good book to read before purchasing solar equipment, or having a solar system installed by a contractor.
Wood Frame House Construction. U.S. Government Printing Office, 1975. S/N 001-000-03528-2.
A basic manual oriented toward low-cost rural building.
Yanda, Bill, and Rick Fisher. *The Food & Heat Producing Solar Greenhouse: Design, Construction and Operation. Rev. ed.* Santa Fe, NM: John Muir Publications, 1979.
A popular book on solar greenhouses by two knowledgeable writers. This revised edition concentrates on heating and growth efficiency.

LAND GRANT UNIVERSITIES IN THE UNITED STATES

By writing to the Agricultural Engineering Extension Office of your state land grant university, plan books and blueprints can be obtained free or for a nominal charge.

Auburn University, Auburn, AL 36830
University of Alaska, Fairbanks, AK 99701
University of Arizona, Tucson, AZ 85721
University of Arkansas, Little Rock, AR 72203
University of California at Davis, Davis, CA 95616
University of California at Riverside, Riverside, CA 92521
Clemson University, Clemson, SC 29631
Colorado State University, Fort Collins, CO 80523
University of Connecticut, Storrs, CT 06268
Cornell University, Ithaca, NY 14853
University of Delaware, Newark, DE 19711
University of Florida, Gainesville, FL 32611
University of Georgia, Athens, GA 30602
University of Hawaii, Honolulu, HI 96822
University of Idaho, Moscow, ID 83843
University of Illinois, Urbana, IL 61801
Iowa State University, Ames, IA 50011

Kansas State University, Manhattan, KS 66505
University of Kentucky, Lexington, KY 40545
Louisiana State University, Baton Rouge, LA 70803
University of Maine, Orono, ME 04473
University of Maryland, College Park, MD 20742
University of Massachusetts, Amherst, MA 01002
Michigan State University, East Lansing, MI 48824
University of Minnesota, St. Paul, MN 55108
Mississippi State University, Mississippi State, MS 39762
University of Missouri, Columbia, MO 65201
Montana State University, Bozeman, MT 59715
University of Nebraska, Lincoln, NB 68503
University of Nevada, Reno, NV 89507
University of New Hampshire, Durham, NH 03824
Purdue University, Lafayette, IN 47907
Rutgers University, New Brunswick, NJ 08903
New Mexico State University, Las Cruces, NM 88003
North Carolina State University, Raleigh, NC 26707
North Dakota State University, Fargo, ND 58102
Ohio State University, Columbus, OH 43210
Oklahoma State University, Stillwater, OK 74074
Oregon State University, Corvallis, OR 97331
Pennsylvania State University, University Park, PA 16802
University of Puerto Rico, Rio Piedras, PR 00928
University of Rhode Island, Kingston, RI 02881
South Dakota State University, Brookings, SD 57006
University of Tennessee, Knoxville, TN 37901
Texas A&M University, College Station, TX 77843
Utah State University, Logan, UT 84322
University of Vermont, Burlington, VT 05401
Virginia Polytechnic Institute and State University,
    Blacksburg, VA 24061
Washington State University, Pullman, WA 99163
West Virginia State University, Morgantown, WV 26506

University of Wisconsin, Madison, WI 53706
University of Wyoming, Laramie, WY 82071

Extension Service, U.S. Department of Agriculture,
Washington, DC 20250

INDEX